VIROLOGY RESEARCH PROGRESS

NOROVIRUSES

OUTBREAKS, CONTROL AND PREVENTION STRATEGIES

VIROLOGY RESEARCH PROGRESS

Additional books in this series can be found on Nova's website under the Series tab.

Additional e-books in this series can be found on Nova's website under the eBooks tab.

VIROLOGY RESEARCH PROGRESS

NOROVIRUSES

OUTBREAKS, CONTROL AND PREVENTION STRATEGIES

JESÚS L. ROMALDE
EDITOR

nova
science publishers
New York

Library of Congress Cataloging-in-Publication Data

ISBN: 978-1-53611-939-8

Library of Congress Control Number: 2017940854

Published by Nova Science Publishers, Inc. † New York

To Mari and Suso

CONTENTS

PREFACE

Norovirus (NoV) are the most common cause of sporadic and epidemic nonbacterial acute gastroenteritis worldwide, with an estimated 21 million cases occurring only in USA annually. NoV disease is characterized by a short duration of the symptoms, which can be severe for people in high risk groups, including children, elderly or immunocompromised patients. In fact, it causes up to 200,000 deaths in children under 5 years of age in developing countries. In addition, after the introduction of rotavirus vaccination programs, NoV are also the most common cause of pediatric gastroenteritis requiring medical care in the high- and middle-income countries. NoV is highly infectious, as even a few particles can cause disease, and infected individuals shed high loads of virus. Transmission occurs primarily through the faecal-oral route, either by consumption of contaminated food or water, or by spreading directly from person to person. Among foods, bivalve shellfish and vegetables are readily vectors due to their mode of consumption raw or slightly cooked. NoV survive for long time on different surfaces and thus, environment may serve as a source of new infections. Therefore, disinfection treatmens of water or foods may help in the prevention of NoV outbreaks. In fact, prevention is currently the key for their control, since no effective treatments or vaccines are available.

All these data are indicative of the importance of noroviruses as human pathogens worldwide, and of the necessity of efforts to develop effective tools for their control and prevention. The purpose of this volume, which presents a compilation of the latest research on these aspects, is to gather together information to help further understand on the prevalence and distribution of genotypes and strains, the potential emergence of new strains of the pathogen, as well as on strategies for the prevention and treatment of Norovirus infections.

The first two chapters provide a state of the art of the NoV epidemiology in two countries, Brazil and Italy, diverse in many aspects, from geographic location to socio-economic development, through climate patterns, population, health systems, etc. M.P. Miogostovich and J.M. Fioretti review NoV data obtained from Brazilian studies, including aspects such as impact in acute gastroenteritis cases, environmental dissemination or genetic diversity demonstrated by molecular epidemiological findings. On the other hand, E. Suffredini and G. La Rosa gather information on NoV epidemiology in the Italian population, and on the prevalence and levels of NoV in water environments, food products and food production environments in Italy.

The third chapter entitled *Norovirus in dogs*, describes the history and epidemiology of canine NoV, from the first descriptions in 2008 to the current worldwide distribution of these specific genotypes. In addition, J. Mesquita and M.S.J. Nascimento present data suggesting that noroviruses may be able to cross the species barrier and raise alerts for potential cross-species transmission events.

In the fourth chapter, C.F. Manso and coworkers describe the results of a study on the evolutionary rates and population dynamics of different genotypes of NoV GII inferred by Bayesian coalescent approach. Estimation of evolutionary rates for minoritary genotypes within genogroup II, such as GII.7, GII.13 and GII.14 are proved for the first time. The fact that NoV GII.14 yields a nucleotide substitution rate and evolutionary dynamic similar to NoV GII.4, suggest the future emergence and widespread epidemics of other genotype variants. Authors claim that such knowledge on the evolutionary dynamics of different NoV genotypes can be helpful in the development of suitable vaccine candidates and other preventing measures targeting the most likely strains circulating in a geographical area.

With numerous norovirus outbreaks, Quantitative Microbial Risk Assessments (QMRAs) for NoV in water and food have emerged with the aim to understand risk of transmission and control of norovirus in a variety of exposure settings. Van Abel and coworkers from the University of Washington, discuss on the different dose-response models proposed by different authors for use in QMRAs for norovirus, their pros and cons or the different variables to be considered. In addition, they identificate the future needs, including determination of the norovirus infectivity, a better understanding of viral aggregation, more low dose human challenge data, studies on genogroup and/or genotype differences in immunity mechanisms, etc., to get a reliable QMRA model.

As mentioned before, ensuring the safety and quality of drinking water is essential in prevention of NoV transmission, especially in developing countries where water scarcity and lack of access to sanitation is common. In the next chapter, *Solar Radiation and Photo-Fentom Systems. Novel Applications for Norovirus Disinfection in Water*, E. Rivadulla and colleagues describe through the review of previous studies together with original research, the feasibility of alternative water treatment processes, such as solar disinfection (SODIS), Fenton and photo-Fenton systems, for the NoV inactivation. Based on the evidences discussed, the authors suggest that SODIS and solar photo-Fenton technologies may constitute feasible alternatives for water treatment in developing countries by providing safer drinking water and making raw or diluted wastewater reuse a long-term future strategy.

David H. Kingsley, in his chapter entitled *High Pressure Processing's Potential to Inactivate Norovirus and other Foodborne Viruses*, discusses research evaluating the potential of high pressure processing (HPP) as an intervention for norovirus contamination of food products. Since foods commonly eaten uncooked such as shellfish and fresh produce can be considered higher risk foods for norovirus transmission, HPP can constitute an effective nonthermal food processing technology against viruses, with the advantage of not detrimentally affect the raw organoleptic properties of oysters or vegetables. The author also discusses on the possibility of variable responses to pressure due to the complexities of food matrices or the diverse resistance of viral genogroups, genotypes and perhaps strains may have. Finally it advise that an intervention such as HPP should be utilized as a fail-safe, or final assurance of food safety, rather than a means to remediate unsanitary foods.

Aptamer has been attracting much attention as a new biomolecular tool, which could be used as highly sensitive and rapid diagnostic methods as well as therapeutic tools. Hirano and coworkers from the Hokkaido University summarize in Chapter 8, entitled *Aptamer: Potential Application to Norovirus Research, Diagnosis, and Therapeutics*, recent developments and applications of norovirus-specific aptamers, including diagnostics, as an alternative for antibodies, or their vast potential to control norovirus infections and mitigate burdens to human society. Some options to focus future research, such as the development of aptasensor platforms towards real-world application, understanding the ability of aptamers to evaluate NoV infectivity or their applications as inhibitors of infections *in vivo* are also discussed.

The last chapter, *Towards an effective norovirus vaccine*, covers one of the aspects that more attention and research efforts have received in the last years, namely the development of a vaccine for preventing NoV infections. The development of vaccines has been hampered by the lack of a realiable cell culture system or an animal model for NoV. As a consequence, candidate formulations are based on recombinant capsid proteins, such as virus-like particles (VLP) and P particles. These candidate NoV vaccines are showing promising results in terms of tolerability and immune response. Future challenges can be the obtention of an appropriate duration of immunity, as well as determination of vaccine regimes, doses, or administration routes. In adition, deeper knowledge on the worldwide circulating strains and on the existence of shared antigenic groups among NoV genotypes as well as the development of models for predicting genotype composition for Nov seasons are indicated as essential to develop a globally effective norovirus vaccine.

I would like to thank all the authors for expending their precious scant time to contribute for this book, for their constancy and tolerance to my frequent pressures regarding deadlines, forms, etc. And last but not least, a special mention of gratitude for the team at Nova Publishers.

Jesús L. Romalde
January 2017

In: Noroviruses
Editor: Jesús L. Romalde
ISBN: 978-1-53611-939-8

Chapter 1

NOROVIRUSES IN BRAZIL

Marize Pereira Miagostovich*
and Julia Monassa Fioretti
Instituto Oswaldo Cruz, Fundação Oswaldo Cruz,
Rio de Janeiro, Brazil

ABSTRACT

The impact of norovirus (NoV) infections in Brazil as well as in the rest of the world was evident in the 2000s, when research laboratories established molecular diagnosis of these viruses associating these emerging viruses with cases of hospitalization by acute gastroenteritis. Currently, studies carried out in different regions in a country of continental dimensions, confirmed the impact of those viruses involved in outbreaks, despite difficulty in reporting cases. Detection of different genotypes of all human genogroups (G) I (GI.1-4, GI.7-8), GII (GII.1-9, GII.12-17, GII.20-22, GII.b, GII.g, GII.e) and IV (GIV.1), animal GII (GII.10, GII.18, GII.19) and variants of GII.4 (US95_96, Kaiso_2003, Asia_2003, Hunter_2004, Yerseke_2006a, Den Haag_2006b, New Orleans_2009, and Sydney_2012 demonstrated its genetic diversity.

* Corresponding Author Address: Laboratório de Virologia Comparada e Ambiental Fundação Oswaldo Cruz - Instituto Oswaldo Cruz 21040-360 Manguinhos - Rio de Janeiro – Brasil; Email: marizepm@ioc.fiocruz.br.

More recently, several recombinant strains, including GII.4 variants were also reported in the country. Despite all data generated in the last decade, Brazil still faces the challenge of establishing methodologies for NoV diagnosis at Public Hospitals, besides demonstrating the importance of active laboratory surveillance for water and food-borne outbreaks where NoV are known as principal agents of infection. This chapter presents an overview of NoV data obtained from Brazilian studies focusing on their impact in acute gastroenteritis cases, environmental dissemination as well as its genetic diversity demonstrated by molecular epidemiological findings.

Keywords: noroviruses, Brazil, genetic diversity

INTRODUCTION

Brazil is the largest country of South and Latin America. The climate in the largest part of the country is tropical and due to its geographical dimensions it features a variety of weather conditions across a large area and varied topography. It is a federation composed of 26 states and the Federal District (FD), Brasília. The states and the FD are grouped into regions that are merely geographical, with no political or administrative divisions, as follows: the North Region [states of Acre (AC), Amazonas (AM), Amapá (AP), Pará (PA), Rondônia (RO), Roraima (RR) and Tocantins (TO)], the Northeast Region [Alagoas (AL), Bahia (BA), Ceará (CE), Maranhão (MA), Paraíba (PB), Pernambuco (PE), Piauí (PI), Rio Grande do Norte (RN) and Sergipe (SE)], the Central-West Region [Goiás (GO), Mato Grosso (MT), Mato Grosso do Sul (MS) and FD], the Southeast Region [Espírito Santo (ES), Minas Gerais (MG), Rio de Janeiro (RJ) and São Paulo (SP)] and the South Region [Paraná (PR), Rio Grande do Sul (RS) and Santa Catarina (SC)]. Many norovirus (NoV) studies carried out in the country were performed in the Southeast and North regions where Reference Laboratories for Gastroenteric viruses are well established. From the 2000th decade, as a result of the establishment of molecular tools for NoV diagnosis in those laboratories, research on those emerging viruses increased and, currently, they are routinely investigated for differential diagnosis of acute gastroenteritis cases.

HISTORY

NoV was first described in Brazil by electron microscopy in stool samples of children with and without diarrhea during the years 1987-1988 in the State of São Paulo (Timenetsky et al. 1993). Later on, Gabbay et al. (1994) conducted the first seroepidemiological study for NoV in children and elderly indigenous in Amazon. In this study, prevalence ranging from 39% to 100% was observed in different communities. In another study, seroprevalence of 71% was demonstrated in a cohort of 135 children aging less than 4 years old in urban shantytown in north-eastern Brazil, by measuring IgG anti-Norwalk like viruses (NLV) by enzyme immunosorbent assay (EIA) with recombinant Norwalk virus capsid protein (Talal et al. 2000).

The first study describing the genetic diversity of NoV demonstrated the occurrence of GI and GII, by nucleotide sequencing of the viral RNA polymerase gene region of NoV, detected in children under one year old hospitalized with severe gastroenteritis and living in poor communities in the city of Fortaleza, State of Ceará (Parks et al. 1999). Five years later, a retrospective study associated NoV with three out of four outbreaks occurred in a day care Centre in Rio de Janeiro, with prevalence ranging from 23 to 67% (Gallimore et al. 2004).

BURDEN OF DISEASE

In the world, as in the Latin American continent, prevalence of NoV was estimated at 18%, regardless of the age of patients (Ahmed et al. 2014). However, higher prevalence can be observed associated to high population densities and/or poor health or socioeconomic conditions (da

Silva Poló et al. 2016). In Brazil, since the second half of the 2000th decade, several studies rapidly demonstrated the increasing impact of NoV in acute gastroenteritis (AGE) cases by assessing NoV in different populations comprising inpatients and outpatients, some of them from outbreaks occurred in close or semi-close settings (Table 1). In the hospital context, wide positivity percentage ranged from some regions such as 8% in the State of Paraná (South) (Raboni et al. 2014) to 63% in the State of Bahia, (Northeast) (Campos et al. 2008). In public hospitals in Rio de Janeiro, NoV infections were associated with 20% of children with acute gastroenteritis cases, 96% of them were caused by GII (Victoria et al. 2007). The impact of NoV infection in pediatric patients revealed similar infection rates or even higher than rotavirus A infection (Borges et al. 2006, Castilho et al. 2006; Soares et al. 2007; Victoria et al. 2007; Andreasi et al. 2008; Nakagomi et al. 2008; Ribeiro et al. 2008; Xavier et al. 2009; Ferreira et al. 2012a, 2012b).

Some studies have also evaluated the presence of NoV in asymptomatic cases demonstrating positivity percentage ranging from 36% to 78% in those cases (Amaral et al. 2015; Barreira et al. 2010; Marques Mendanha de Oliveira et al. 2014; Castilho et al. 2006; Vicentini et al. 2013). NoV was also observed in 26.7% of cases of persistent diarrhea in children under three years of age treated in outpatient clinics from different hospitals in the state of São Paulo (Castilho et al. 2006).

Data from several studies conducted in the country discloses that NoV seasonality in Brazil is still contradictory. In a six-year study (1998-2005) conducted in Rio de Janeiro (Southeast), Soares et al. (2007) observed no seasonality pattern. However, other studies performed in the same region of the country (Rio de Janeiro and Espírito Santo) evidenced seasonality in dry months (Victoria et al. 2007, Ribeiro et al. 2008; Ferreira et al. 2010a). In contrast, other Brazilian studies found higher detection in periods of high humidity (Borges et al. 2006; Ferreira et al. 2010b). Recently, Fumian et al. (2016a) observed a variation rate in NoV detection, being 33% of studied cases during the period between October/2013 - February/2014 and 13.8% during September - December/2014 in clinical samples collected from different Brazilian states. In this study, an increase of NoV incidence during the summer season was observed.

Genetic Diversity, Temporal Dynamics of Norovirus GII.4 Variants and Emergence of New Variants

Partial nucleotide sequencing of the genome including RNA dependent RNA polymerase (RdRp), known as region B, and different regions encoding the viral capsid (regions C and D) have been used on researches to characterize NoV strains in Brazil (Kojima et al. 2002; Beuret et al. 2002; Vinjé et al. 2004). Molecular characterization using those protocols revealed the great diversity of human NoV circulating in the country including 6 genotypes belonging to GI, 21 GII and 1 GIV. In addition, diversity is provided by current GII.4 variants, as well as the large number of recombination strains reported (Figure 1).

GII.4 NoV is the most important genotype associated with outbreaks and sporadic cases in the country, detected in 78% of all samples in a study performed in 13 Brazilian states (Fioretti et al. 2011). In the Middle Paraíba region (Southeast) this genotype was also detected in 66% of cases affecting different age groups during an outbreak occurred in 2005 (Ferreira et al. 2008). However, a 15-year retrospective study, including acute gastroenteritis cases from the 1990th decade, identified the occurrence of outstanding genotypes not reported in the country and a broader monthly distribution of NoV in 33.4% of samples (151/452) (Ferreira et al. 2012a). In another eight year-retrospective study (2004-2011) that investigated 2,265 stool samples from AGE outbreaks in the state of Rio Grande do Sul (South), NoV was detected in 817 stool samples (36.1%) associated with 327 outbreaks (44.1%), and about 70% of them resulted from genotype GII.4 and its variants (de Andrade et al., 2014).

Table 1. Occurrence of Noroviruses in stool samples in Brazil

Region	State	Population studied	Age group (years)	Study Period	Samples analysed (% Positivity)	Genogroup/ Genotypes	Reference
Northeast	Maranhão	Hospitalized	< 3	1997 - 1999	136 (17.6)	GII.3, GII.4, GII.5 GII.7, GII.12, GII.16	Portal et al., 2016
	Ceará	Hospitalized	< 10 years	2008 - 2009	200 (17)	GII.3, GII.4 GII.6, GII.12	Sá et al., 2015
	Pernambuco	Hospitalized and Ambulatory	< 5	2004 - 2005	233 (15)	NR	Nakagomi et al., 2008
	Bahia	Hospitalized	Children and adults	2006	127 (63)	GII.3, GII.4, GII.9	Campos et al., 2008
North	Pará	Hospitalized	< 3	2008 - 2009	92 (62)	NR	Siqueira et al., 2011
		Quilombola community	< 10	2008 - 2010	81 (19.7)	GI.2, GII.4, GII.6 GII.7, GII.17, GII.20	Aragão et al., 2013
		Hospitalized	< 6	2012	85 (34.1)	GII.4 GII.7	Fumian et al., 2013b
		Hospitalized	< 5	2008 - 2011	483 (35.4)	NR	Siqueira et al., 2013a
		Hospitalized	< 5	2008 - 2010	348 (36.5)	GII.4 GII.b	Siqueira et al., 2013b
	Rondônia	Hospitalized	< 6 years	2010 - 2012	591 (7.8)	NR	Amaral et al., 2015
	Acre	Hospitalized	< 8	2012	25 (48)	GII.4	Silva et al., 2013
Southeast	São Paulo	Hospitalized and Ambulatory	< 3	1995 - 1999	234 (33.3)	GI.3, GI.4, GII.2 GII.3, GII.4, GII.6	Castilho et al., 2006
		Hospitals Health Care	NR	1995, 2004 - 2006	204 (15.7)	GII	Morillo et al., 2008
		Day care Facility	21 - 98	2005	9 (44.4)	GII	Menezes et al., 2010
		Children's day care	Children	2010 - 2011	100 (2)	NR	Castro et al., 2015a
		NR	NR	2010	327 (32.7)	GI, GII	Morillo et al., 2011a
		Correctional facilities	NR	2010	27 (29.6)	GII	Luchs et al., 2011
		Hospitalized	< 5	2004 - 2009	89 (29.2)	GI GII	Cilli et al., 2011
	Rio de Janeiro	Children's day care	< 5	1996 - 1998	132 (31)	GII.3, GII.4	Gallimore et al., 2004
		Hospitalized	Children	2004	318 (20)	GI.2, GII.2, GII.4	Victoria et al., 2007
		Hospitalized and Ambulatory	< 10	1998 - 2005	289 (14.5)	GI GII	Soares et al., 2007
		Ambulatory		2005	50 (66)	GII.4	Ferreira et al., 2008

Region	State	Population studied	Age group (years)	Study Period	Samples analysed (% Positivity)	Genogroup/ Genotypes	Reference
		Hospitalized and Ambulatory	< 1 > 50	2005 - 2008	1087 (35.1)	GI.2, GI.3, GII.3 GII.4, GII.6, GII.7 GII.8, GII.14	Ferreira et al., 2010
		Children's day care	< 5	1994 - 2008	452 (33.4)	GI.2, GI.3, GI.8 GII.2, GII.3, GII.4 GII.6, GII.7, GII.12 GII.17	Ferreira et al., 2012a
		Hospitalized	< 2	2013	79 (30.3)	GII.4, GII.6, GII.21	Ferreira et al., 2012b
	Espírito Santo	Hospitalized and Ambulatory	< 3	2003 - 2004	319 (16.3)	GII.3, GII.4, GII.6 GII.8, GII.14	Barreira et al., 2010
		Quilombola community	< 11	2007 - 2009	397 (4)	GI.1, GI.3, GI.4, GII.4, GII.6, GII.12	Vicentini et al., 2013
		Hospitalized	< 12	2004 - 2006	68 (39.7)	NR	Ribeiro et al., 2008
South	Rio Grande do Sul	NR	NR	2006 - 2007	365 (13.2)	GII	Georgiadis et al., 2010
		Ambulatory	<1 - > 50 years	2004 - 2011	2265 (36.1)	GI.1, GI.3, GII.2 GII.3, GII.4, GII.6 GII.12, GII.13, GII.14 GII.15, GII.17, GII.21	De Andrade et al., 2014
	Paraná	Hospitalized	Children	2010 - 2011	225 (8)	GII.4	Raboni et al., 2014
Central-West	Goiás	Children's day care	< 5	2009 - 2011	539 (8)	GI.1, GI.7, GII.1, GII.2, GII.4, GII.6	Marques Mendanha de Oliveira et al., 2014
-	Seven states *	Hospitals Health Cares	< 5 to > 60	2006 - 2009	245 (67.3)	GI GII	Morillo et al., 2011b
-	Thirteen states **	NR	< 51	2005 - 2009	64 (NR)	GII.4, GII.6, GII.7, GII.12, GII.19, GII.17	Fioretti et al., 2011

NR: Not reported

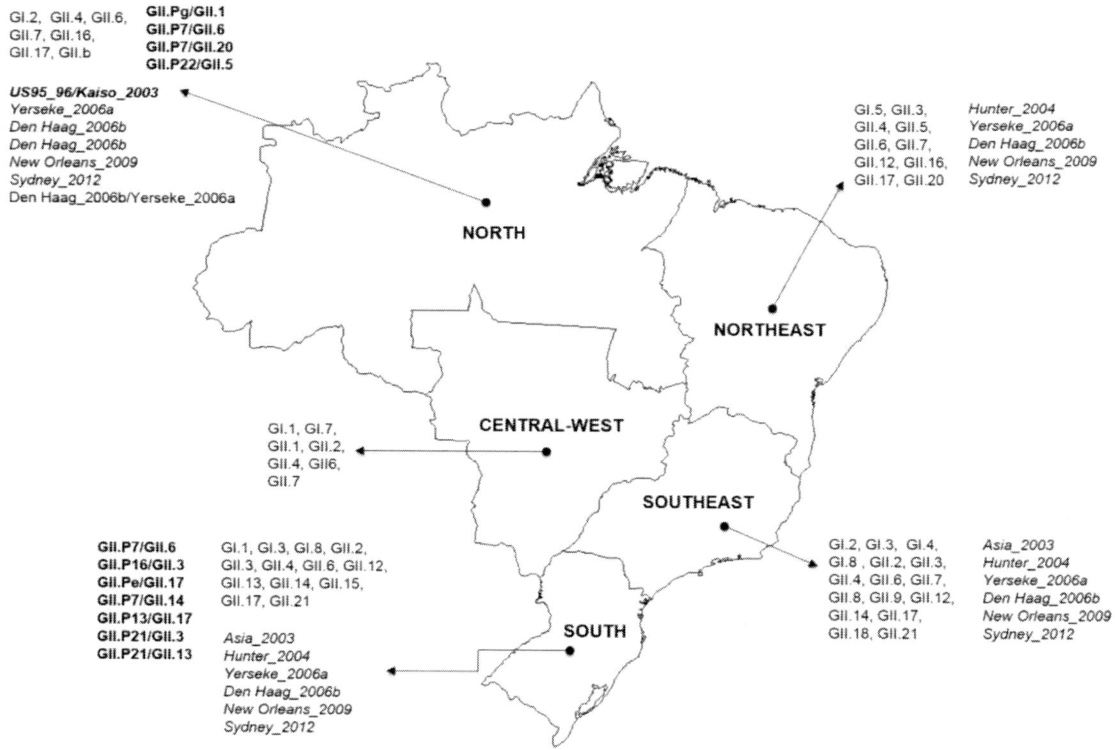

Figure 1. Distribution of noroviruses genotypes in Brazil. Genotypes in bold represent recombinant viruses and in italic represent GII.4 variants.

Table 2. Occurrence of Noroviruses in environmental samples

Region	State	Type of sample	Study Period	Samples analysed (% Positivity)	Genogroup/ Genotypes	Concentration (GC/L)	Reference
North	Manaus	Stream	2004 - 2005	52 (5.8)	GII	NR	Miagostovich et al. 2008
		River and Stream	2011 - 2012	212 (7.4)	GII	4.8E +02 - 4.8E +04	Vieira et al. 2016
	Belém	River	2008 - 2010	24 (4.2)	GIV.1	NR	Teixeira et al. 2016
		Bay	2008 - 2010	24 (4.2)	GIV.1	NR	
		Stream	2008 - 2010	24 (12.5)	GIV.1	NR	
Southeast	Rio de Janeiro	Biosolids	2011	11 (45.5)	GII	5.2E+02 - 1.2E+02	Prado et al., 2013
		Sludge	2011	12 (50)	GII	1.6E+04 - 4.9E+05	Prado et al., 2014
		Beach	2012	74 (14)	GII	5.8E+02 - 1.6E+04	Victoria et al., 2014
		Wastewater	2009 - 2010	24 (55)	GII.4, GII.g	1.5E+04 -.03E+05	Fumian et al., 2013
		Wastewater	2005	48 (58)	GI.2., GII.3, GII.4, GII.b	6.4E+02 - 7.3E+03	Victoria et al., 2010a
		Wastewater	2005 - 2008	34 (38.2)	GII.4	3.2E+01 - 1.1E+04	Prado et al., 2011
		Lagoon	2007 - 2008	120 (17.5)	GII	NR	Vieira et al. 2012
		River	NR	6 (33.3)	GII	6.8E+03 - 8.6E+03	Calgua et al. 2013
			2007 - 2008	12 (33.3)	GII	NR	Vieira et al. 2012
			2008 - 2009	108 (9.3)	GI, GII	NR	Miagostovich et al. 2014
	Espirito Santo	Mangrove	2008 - 2009	42 (4.8)	GII	NR	Keller et al., 2013
South	Florianópolis	Lagoon	2007 - 2008	12 (8.3)	GI	NR	Victoria et al. 2010b
		Creek	2007 - 2008	12 (71.7)	GI.2, GII.2, GII.4	1.14E+02 - 3.7E+03	
		Beach	2009 - 2010	132 (12)	GI, GII	E+02 - E+05	Moresco et al., 2012

GC/L: (genome copies/L)
NR: Not reported

Bayesian inference of GII.4 NoV obtained from hospitalized patients with acute gastroenteritis in the state of Rio de Janeiro during the period 2004-2007 revealed that the viral population of NoV in Rio de Janeiro fitted the expansion growth model. This study identified the year of 2002 as the time of the most recent common ancestor (T_{MRCA}) of all GII.4 variants analysed (Victoria et al. 2009a). Silva et al. (2013) described for the first time in Brazil, circulation of GII.4 variant Sydney_2012 in fecal samples collected from children in a study performed in Rio Branco, State of Acre, in northern Brazil. Later, temporal dynamics of GII.4 NoV was assessed in a nine-year study (2004-2012) that analyzed 147 virus strains selected from 12 states from the three most populous geographic regions of the country (southeast, south and northeast). Data revealed circulation of six GII.4 variants such as: Asia_2003, Hunter_2004, Yerseke_2006a, Den Haag_2006b, New Orleans_2009 and Sydney_2012 with clear emergence and replacement patterns over the years studied (Fioretti et al. 2014). The mean evolutionary rate of subdomain P2 (7.3×10^{-3} subst./site/year) from GII.4 NoV Brazilian strains was similar to rates reported in previous studies that evaluated the entire gene coding for VP1 ($1.9 - 9.0 \times 10^{-3}$ subst./site/year) (Bok et al. 2009; Bull et al. 2010; Siebenga et al. 2010; Boon et al. 2011). Analysis of the T_{MRCA} of these GII.4 Brazilian strains demonstrated that the diversification process started at least one year before the variant became a dominant strain and caused epidemics, corroborating a study that suggests that a strain circulates at low levels accumulating mutations for some time before becoming pandemic (Eden et al. 2010). In this Brazilian study, the first variant detected was Asia_2003 that was predominant in 2004, later being replaced by Hunter_2004 in 2005. In 2006, Den Haag_2006b variant emerged, replacing the previous predominant variant, and remained as the most prevalent during the two following years (2007 and 2008). In 2007, Yerseke_2006a variant was first detected however, it did not replace the previous variant Den Haag_2006b, though both variants co-circulated at equal proportions during 2008-2009. Detection rate of Yerseke_2006a decreased by 2010; while Den Haag_2006b remained detectable until 2011. The emergence of the New Orleans_2009 variant was observed in 2009, and from 2010 to 2011 this variant became predominant, being replaced in 2012 by the Sidney_2012 (Fioretti et al. 2014). By reconstruction of the demographic history of the Brazilian clade Den Haag_2006b this study also demonstrated, that after a rapid increase in 2005, the population reached an effective maximum size in early 2006 (when first detected) and then declined until mid-2008. Since then, this variant increased sharply again, reaching a new maximum effective population size in 2009. An Argentinean study that evaluated the temporal dynamics of this variant observed that the effective size of the population increased in mid-2004 and began to decline in 2008 (Fernández et al. 2012). Such complex demographic patterns of this variation may be related to the existence of two distinct Den Haag_2006b sublineages in Brazil: sublineage "Y" (younger), which emerged around 2008, is an ancestor of sublineage "O" (older) which probably arose in 2005. The emergence of sublineage "Y" coincides with the second expansion of Den Haag_2006b phase. Brazilian sublineage "Y" grouped into the same clade as Taiwan, Sweden and Japan strains. It is noteworthy that strains of sublineage "Y" showed a large number of substitutions at residues 255, 357, 368 and 425, when compared to subgroup "O" (Fioretti et al. 2014). Residue 368 is located in the putative epitope "A" of viral capsid, associated as the most likely responsible for the difference in antigenicity GII.4 variants (Debbink et al. 2012; Lindesmith et al. 2012), suggesting that "O" and "Y" Den Haag_2006b sublineages may represent two distinct antigenic NoV GII.4 variants.

Another seven-year study (2004-2011) conducted in the State of Rio Grande do Sul (Southern), also using specific primers for amplification of P2 subdomain (Vega et al. 2011), demonstrated circulation of Asia_2003, Hunter_2004, Yerseke_2006a, Den Haag_2006b and New Orleans_2009 variants, being variant Den Haag_2006b, the most widely distributed, found in seven mesoregion states and detected for four consecutive years (2006-2009) (de Andrade et al. 2014).

By the year 2016, no other GII.4 variant was described after the emergence of Sydney_2012. However, in the State of Pará, recombination among different GII.4 variants US95_96/Kaiso_2003 and Den Haag_2006b/Yerseke_2006a were observed in infected hospitalized children with severe acute gastroenteritis in 2003 and in 2009, respectively (Siqueira et al. 2016).

Antigenic drift and recombination of hotspot, mainly of ORF1/ORF2 junction region was reported as an important mechanism for NoV evolution leading to emergence of novel viruses (White 2014; Bull et al. 2005; 2007; Chhabra et al. 2010; Mahar et al. 2011; Eden et al. 2013; Arana et al. 2014). In the last five years, several recombinant NoV strains were described in three studies conducted in north and southeast Brazil. Using SimPlot and Bootscan analysis of recombination breakpoints near ORF1/2 junction, a GII.P7/GII.20 recombinant strain was demonstrated in Pará (North) from an African-descendant community while uncommon recombinants such as GII.P22/GII.5, GII.P7/GII.6 and GII.Pg/GII.1 was described in Manaus (Fumian et al. 2012; Hernandez et al. 2016). NoV recombinant strains characterized as GII.P7/GII.6; GIIPg/GII.12; GII.P16/GII.3; GII.e/GII.17; GII.P7/GII.14; GII.P13/GII.17; GII.P21/GII.3 and GII.P21/GII.13 were also identified in 23 out of 27 (85%) non-GII.4 strains associated to outbreaks in Southern Brazil (Fumian et al. 2016b).

Recently, the main NoV outbreaks occurred in parts of Asia were linked to newly emerged GII.P17_GII.17 strain, which could replace previous circulating Sydney_2012 GII.4 variant (Gao et al. 2015). Although this genotype was also detected in cases of acute gastroenteritis in Europe, United States and Australia, replacement of Sydney_2012 was not observed in these regions (de Graaf et al. 2015). If this pattern is confirmed, it is possible that GII.17 becomes the main genotype associated with acute gastroenteritis outbreaks.

Data on Environmental and Food Occurrence

In Brazil, the first study reporting NoV dissemination in the environment was conducted in 2004 when presence of gastroenteric viruses in sewage contaminated streams of Manaus was assessed (Miagostovich et al. 2008). Later on, environmental studies determined occurrence of NoV in several matrixes including raw sewage from wastewater treatment plants (WWTP), where virus concentration could reach 1.5×10^4-0.03×10^5 genome copies (GC)/L (Fumian et al. 2013a). The distribution of NoV in environmental samples is highly variable and dependent on the matrix studied, the community's epidemiological profile and geographical conditions (Table 2). NoV were detected from wastewater and sewage sludge samples from WTTPs using conventional activated sludge treatment after anaerobic digestion (Prado et al. 2011, 2013, 2014), in marine recreational waters in beaches of Florianópolis (Victoria et al. 2010b; Moresco et al. 2012) and Rio de Janeiro (Vieira et al. 2012; Victoria et al. 2014). A study performed during June/2007 to May/2008 assessed NoV contamination in aquatic ecosystems in the city of Florianópolis, demonstrating its detection in treated sewage, superficial (creek and brackish lagoon) and drinking water. The co-circulation of three genotypes (GII.2, GII.4 and GI.3) was also verified (Victoria et al. 2010b).

In Brazil, more than 10,000 outbreaks associated with food contamination were reported in recent years; however, over half of them do not present a defined etiological agent. Despite limited data, NoV was associated with 38 food-borne outbreaks (0.9%) in a total of 9,719 cases reported by the Ministry of Health 2000-2014 (ANVISA, 2014). This association is generally established by analysis of clinical samples from gastroenteric cases involved in those outbreaks.

Brazilian standards for microbiological food control do not include virus as parameters for food products, mainly due to the lack of methods available in microbiology laboratories for NoV recovering from food matrixes (ANVISA, 2001). Nevertheless, due to the increasing epidemiological importance of NoV foodborne outbreaks, especially on cruise ships, the National Program for Strengthening Health Surveillance that acts in Ports, Airports and Borders published, on December 6th 2012, the Ordinance no. 2795 establishing NoV diagnosis in Central Laboratories of the states (Amazonas, Bahia, Ceará, Pará, Pernambuco, Rio de Janeiro, Santa Catarina and São Paulo) in the route of the cruise season. However, the direct detection of viruses from food samples naturally contaminated in the country is still currently performed by research laboratories that are evaluating methods for virus recovery from different environmental and food matrixes (Gouvea et al.1994; Victoria et al. 2009b; Fumian et al.

2009; Morillo et al. 2012; Brandão et al. 2014a,b; de Abreu Corrêa 2012a,b; de Abreu Corrêa &, Miagostovich 2013; Ganime et al. 2015; Garcia et al. 2015; Melgaço et al. 2016).

In 2010, NoV GI.1 and GII.4 were identified from herbed butter, cheese and white sauce samples served on a cruise ship, indicating these products as sources of acute gastroenteritis outbreak in a cruise ship in the country (Morillo et al. 2012). NoV were also detected in *Crassostera gigas* cultivated in marine farms with concentrations ranging from 1.9 x 10^5 to 7.5 x 10^{13} GC/g (Souza et al. 2012) and in cheese samples acquired in markets in Rio de Janeiro (Melgaço, 2016).

Data on Animal Norovirus

In a study conducted with stool samples and intestinal contents of swine from pig's herds in Rio de Janeiro, porcine NoV (PoNoV) was identified in 20 out of 46 samples (20%). Nucleotide sequencing of both RdRp and VP1 regions identified GII.11, GII.18, GII.19 and a possible GII.9 and GII.11 recombinant, supporting PoNoV circulation in Brazilian pig herds (Cunha et al. 2010a, 2010b). Investigation of a natural PoNoV infection from adult animals from an important Brazilian pig-production region also reported GII.11 from asymptomatic pigs aged 9 to 24 weeks old from grower-to-finish herds located in Paraná state, Brazilian Southern region (Silva et al. 2015). Recently, NoV GIV was also detected at 5.98×10^5 GC/g concentration from kittens stool samples obtained in rescue shelters, being the first report of enteric FeNoV in diarrheic cats in South America (Castro et al. 2015b).

Future Challenges

Besides NoV researches reported in this chapter, some Brazilian studies have also evaluated methodologies for NoV diagnosis as immunosorbent assay (EIA) commercially available or even protocol adjustments made in-house to ensure more than one diagnostic methodology to be made available (Morillo et al. 2011b; Siqueira et al. 2011; Lamounier et al. 2015; Fumian et al. 2016a). Thus, despite efforts to demonstrate the impact of NoV infections in Brazil, health authorities still face a challenge to establish a routine surveillance system to investigate prevalence of those viruses throughout different regions. A network of sentinel laboratories should be encouraged to share epidemiological and molecular information such as in developed countries. Laboratorial and epidemiological data on circulation of genotypes and variants are decisive tools to a successful active surveillance and to estimate the impact of rapid evolution on the health of population.

Competing Interest

The authors inform that there are no conflicts of interest. The use of trade names in this publication does not endorse any commercial products.

Acknowledgments

The authors thanks the editors and the institutions that have supported projects in the Environmental Virology field such as Oswaldo Cruz Institute/Fiocruz, the Brazilian National Council for Scientific and Technological Development (CNPq) and Carlos Chagas Filho Foundation for Research Support in the State of Rio de Janeiro (Faperj). MPM (Research Productivity) is a fellowship of CNPq. This chapter is under the scope of the activities of Fiocruz as a collaborating Centre of PAHO/WHO of Public and Environmental Health.

REFERENCES

Ahmed SM, Hall AJ, Robinson AE, Verhoef L, Premkumar P, Parashar UD, Koopmans M, Lopman BA. 2014. Global prevalence of norovirus in cases of gastroenteritis: a systematic review and meta-analysis. *Lancet Infect. Dis.* 14(8):725-30.

Amaral MS, Estevam GK, Penatti M, Lafontaine R, Lima IC, Spada PK, Gabbay YB, Matos NB. 2015. The prevalence of norovirus, astrovirus and adenovirus infections among hospitalised children with acute gastroenteritis in Porto Velho, state of Rondônia, western Brazilian Amazon. *Mem. Inst. Oswaldo Cruz* 110: 215-221.

Andreasi MS, Cardoso D, Fernandes SM, Tozetti IA, Borges AM, Fiaccadori FS, Santos RA, Souza M. 2008. Adenovirus, calicivirus and astrovirus detection in fecal samples of hospitalized children with acute gastroenteritis from Campo Grande, MS, Brazil. *Mem. Inst. Oswaldo Cruz* 103: 741-744.

Anvisa - Agência Nacional de Vigilância Sanitária [National Health Surveillance Agency]. 2001. Ministério da Saúde. Resolução RDC nº12, de 2 de janeiro de 2001. Aprova regulamento Técnico sobre Padrões Microbiológicos para alimentos e seus anexos I e II. *Diário Oficial da União, Poder Executivo* [Ministry of Health Resolution RDC No. 12 of January 2, 2001. Approves a Technical Regulation on Microbiological Standards for Food and its Annexes I and II. Official Journal of the Union, Executive Branch], Brasília-DF, Brasil, n.7-E, 10 Jan. 2001. Seção1, pag 45. http://bvsms.saude.gov.br/bvs/saudelegis/anvisa/2001/ res0012_02_01_2001.html. Accessed 14 Jul. 2015.

Anvisa - Agência Nacional de Vigilância Sanitária. [National Health Surveillance Agency] 2014. Ministério da Saúde. Programa de Análise de Resíduos de Agrotóxicos em Alimentos (PARA), *Relatório Complementar,* 2012 [Ministry of Health. Program for the Analysis of Agrochemical Waste in Foods (PARA), Complementary Report, 2012]. Brasília-DF, Brasil, outubro 2014. http://portal.

anvisa.gov.br/wps/por tal/anvisa/anvisa/home/agrotoxicotoxicologia. Accessed 14 Jul. 2015.

Aragão GC, Mascarenhas JD, Kaiano JH, de Lucena MS, Siqueira JA, Fumian TM, Hernandez J D, de Oliveira CS et al. 2013. Norovirus diversity in diarrheic children from naAfrican-descendant settlement in Belém, Northern Brazil. *PLoS One* 8: e56608.

Arana A, Cilla G, Montes M, Gomariz M, Pérez-Trallero E. 2014. Genotypes, recombinant forms, and variants of norovirus GII.4 in Gipuzkoa (Basque Country, Spain), 2009-2012. *PLoS One* 9: e98875.

Barreira DM, Ferreira MS, Fumian TM, Checon R, de Sadovsky AD, Leite JP, Miagostovich MP, Spano LC. 2010. Viral load and genotypes of noroviruses in symptomatic and asymptomatic children in Southeastern Brazil. *J. Clin. Virol.* 47: 60-64.

Beuret C, Kohler D, Baumgartner A, Lüthi TM. 2002. Norwalk-like virus sequences in mineral waters: one-year monitoring of three brands. *Appl. Environ. Microbiol.* 68: 19251931.

Bok K, Abente EJ, Realpe-Quintero M, Mitra T, Sosnovtsev SV, Kapikian AZ, Green KY. 2009. Evolutionary dynamics of GII.4 noroviruses over a 34-year period. *J. Virol.* 83: 11890-11901.

Boon D, Mahar JE, Abente EJ, Kirkwood CD, Purcell RH, Kapikian AZ, Green KY, Bok K. 2011. Comparative evolution of GII.3 and GII.4 norovirus over a 31-year period. *J. Virol.* 85: 8656-8666.

Borges AM, Teixeira JM, Costa PS, Giugliano LG, Fiaccadori FS, Franco Rde C, Brito WM, Leite JP et al. 2006. Detection of calicivirus from fecal samples from children with acute gastroenteritis in the West Central region of Brazil. *Mem. Inst. Oswaldo Cruz* 101: 721724.

Brandão MLL, Almeida DO, Marin VA, Miagostovich, MP. 2014a. Recovery of Norovirus from lettuce (*Lactuca sativa*) using an adsorption-elution method with a negatively charged membrane: comparison of two elution buffers. *Visa em Debate* 2: 58-63.

Brandão MLL, Almeida DO, Bispo FCP, Bricio SML, Marin VA, Miagostovich, MP. 2014b. Assessment of microbiological contamination of fresh, minimally processed, and ready-toeat lettuces (*Lactuca sativa*), Rio de Janeiro State, *Brazil. J Food Sci.* 79: M961-M966.

Bull RA, Hansman GS, Clancy LE, Tanaka MM, Rawlinson WD, White PA. 2005. Norovirus recombination in ORF1/ORF2 overlap. *Emerg. Infect. Dis.* 11: 1079-1085.

Bull RA, Tanaka MM, White PA. 2007. Norovirus recombination. *J. Gen. Virol.* 88: 33473359.

Bull RA, Eden JS, Rawlinson WD, White PA. 2010. Rapid evolution of pandemic noroviruses of the GII.4 lineage. *PLoS Pathog.* 26: 6: e1000831.

Calgua B, Fumian T, Rusiñol M, Rodriguez-Manzano J, Mbayed VA, Bofill-Mas S, Miagostovich M, Girones R. 2013. Detection and quantification of classic and emerging viruses by skimmed-milk flocculation and PCR in river water from two geographical areas. *Water Res.* 47: 2797-2810.

Campos GS, Moreau VH, Bandeira A, Barberino G, Almeida PF, Amador DM, de Lima MO, Sardi SI. 2008. Molecular detection and genetic diversity of norovirus in hospitalized young adults with acute gastroenteritis in Bahia, Brazil. *Arch. Virol.* 153: 1125-1129.

Castilho JG, Munford V, Resque HR, Fagundes-Neto U, Vinjé J, Rácz ML. 2006. Genetic diversity of norovirus among children with gastroenteritis in São Paulo State, Brazil. *J. Clin. Microbiol.* 44: 3947-3953.

Castro ED, Germini MC, Mascarenhas JD, Gabbay YB, de Lima IC, Lobo P dos S, Fraga VD, Conceição LM et al. 2015a. Enteropathogens detected in a daycare center, Southeastern Brazil: bacteria, virus, and parasite research. *Rev. Inst. Med. Trop. Sao Paulo* 57: 27-32.

Castro TX, Cubel Garcia R de C, Fumian TM, Costa EM, Mello R, White PA, Leite JP. 2015b. Detection and molecular characterization of caliciviruses (vesivirus and norovirus) in an outbreak of acute diarrhea in kittens from Brazil. *Vet. J.* 206: 115-117.

Cilli A, Luchs A, Morillo SG, Costa FF, Carmona R de C, Timenetsky M do C. 2011. Characterization of rotavirus and norovirus strains: a 6-year study (2004-2009). *J. Pediatr.* (Rio J) 87: 445-449.

Chhabra P, Walimbe AM, Chitambar SD. 2010. Molecular characterization of three novel intergenotype norovirus GII recombinant strains from western India. *Virus Res.* 147: 24246.

Cunha JB, de Mendonça MC, Miagostovich MP, Leite JP. 2010a. Genetic diversity of porcine enteric caliciviruses in pigs raised in Rio de Janeiro State, Brazil. *Arch. Virol.* 155: 13011305.

Cunha JB, de Mendonça MC, Miagostovich MP, Leite JP. 2010b. First detection of porcine norovirus GII.18 in Latin America. 89: 126-129.

da Silva Poló T, Peiró JR, Mendes LC, Ludwig LF, de Oliveira-Filho EF, Bucardo F, Huynen P, Melin P et al. 2016. Human norovirus infection in Latin America. *J. Clin. Virol.* 78: 111119.

de Abreu Corrêa A, Carratala A, Barardi CR, Calvo M, Girones R, Bofill-Mas S. 2012a. Comparative inactivation of murine norovirus, human adenovirus, and human JC polyomavirus by chlorine in seawater. *Appl. Environ. Microbiol.* 78: 6450-6457.

de Abreu Corrêa A, Souza DS, Moresco V, Kleemann CR, Garcia LA, Barardi CR. 2012b. Stability of human enteric viruses in seawater samples from mollusc depuration tanks coupled with ultraviolet irradiation. *J. Appl. Microbiol.* 113: 1554-1563.

de Abreu Corrêa A. and Miagostovich MP. 2013. Optimization of an Adsorption–Elution Method with a Negatively Charged Membrane to Recover Norovírus from Lettuce. *Food Environ. Virol.* 5: 144-149.

de Andrade Jda S, Rocha MS, Carvalho-Costa FA, Fioretti JM, Xavier Mda P, Nunes ZM, Cardoso J, Fialho AM et al. 2014. Noroviruses associated with outbreaks of acute gastroenteritis in the State of Rio Grande do Sul, Brazil, 2004-2011. *J. Clin. Virol.* 61: 345352.

de Graaf M, van Beek J, Vennema H, Podkolzin AT, Hewitt J, Bucardo F, Templeton K, Mans J et al. 2015. Emergence of a novel GII.17 norovirus - End of the GII.4 era? *Euro Surveill.* 20. pii: 21178.

Eden JS, Bull RA, Tu E, McIver CJ, Lyon MJ, Marshall JA, Smith DW, Musto J et al. 2010. Norovirus GII.4 variant 2006b caused epidemics of acute gastroenteritis in Australia during 2007 and 2008. *J. Clin. Virol.* 49: 265-271.

Eden JS, Tanaka MM, Boni MF, Rawlinson WD, White PA. 2013. Recombination within the pandemic norovirus GII.4 lineage. *J. Virol.* 87: 6270-6282.

Fernández MD, Torres C, Poma HR, Riviello-López G, Martínez LC, Cisterna DM, Rajal VB, Nates SV et al. 2012. Environmental surveillance of norovirus in Argentina revealed distinct viral diversity patterns, seasonality and spatio-temporal diffusion processes. *Sci. Total Environ.* 437: 262-269.

Ferreira MS, Xavier MP, Fumian TM, Victoria M, Oliveira SA, Pena LH, Leite JP, Miagostovich MP. 2008. Acute gastroenteritis cases associated with noroviruses infection in the state of Rio de Janeiro. *J. Med. Virol.* 80: 338-344.

Ferreira MS, Victoria M, Carvalho-Costa FA, Vieira CB, Xavier MP, Fioretti JM, Andrade J, Volotão EM et al. 2010. Surveillance of norovirus infections in the state of Rio De Janeiro, Brazil 2005-2008. *J. Med. Virol.* 82: 1442-1448.

Ferreira MS, Xavier M da P, Tinga AC, Rose TL, Fumian TM, Fialho AM, de Assis RM, Carvalho Costa FA et al. 2012a. Assessment of gastroenteric viruses frequency in a children's day care center in Rio De Janeiro, Brazil: a fifteen year study (1994-2008). *PLoS One* 7: e33754.

Ferreira MS, Cubel Garcia R de C, Xavier M da P, Ribeiro RL, Assis RM, Mota M do C, Leite JP, Miagostovich MP et al. 2012b. Genotyping of gastroenteric viruses in hospitalised children: first report of norovirus GII.21 in Brazil. *Mem. Inst. Oswaldo Cruz* 107: 10641067.

Fioretti JM, Ferreira MS, Victoria M, Vieira CB, Xavier Mda P, Leite JP, Miagostovich MP. 2011. Genetic diversity of noroviruses in Brazil. *Mem. Inst. Oswaldo Cruz* 106: 942-947.

Fioretti JM, Bello G, Rocha MS, Victoria M, Leite JP, Miagostovich MP. 2014. Temporal dynamics of norovirus GII.4 variants in Brazil between 2004 and 2012. *PLoS One* 9: e92988.

Fumian TM, Leite JP, Marin VA, Miagostovich MP. 2009. A rapid procedure for detecting noroviruses from cheese and fresh lettuce. *J. Virol. Methods* 155: 39-43.

Fumian TM, Aragão GC, Mascarenhas JD, Kaiano JH, Siqueira JA, Soares LS, Linhares AC, Gabbay YB. 2012. Detection of a novel recombinant strain of norovirus in an Africandescendant community from the Amazon region of Brazil in 2008. *Arch. Virol.* 157: 23892392.

Fumian TM, Vieira CB, Leite JP, Miagostovich MP. 2013a. Assessment of burden of virus agents in an urban sewage treatment plant in Rio de Janeiro, Brazil. *J. Water Health.* 11: 110-119.

Fumian TM, Justino MC, D'Arc Pereira Mascarenhas J, Reymão TK, Abreu E, Soares L, Linhares AC, Gabbay YB. 2013b. Quantitative and molecular analysis of noroviruses RNA in blood from children hospitalized for acute gastroenteritis in Belém, Brazil. *J. Clin. Virol.* 58: 31-35.

Fumian TM, Leite JP, Rocha MS, de Andrade JS, Fioretti JM, de Assis RM, Assis MR, Fialho AM et al. 2016a. Performance of a one-step quantitative duplex RT-PCR for detection of rotavirus A and noroviruses GII during two periods of high viral circulation. *J. Virol. Methods* 228: 123-129.

Fumian TM, da Silva Ribeiro de Andrade J, Leite JP, Miagostovich MP. 2016b. Norovirus Recombinant Strains Isolated from Gastroenteritis Outbreaks in Southern Brazil, 20042011. *PLoS One* 11: e0145391.

Gabbay YB, Glass RI, Monroe SS, Carcamo C, Estes MK, Mascarenhas JD, Linhares AC. 1994. Prevalence of antibodies to Norwalk virus among Amerindians in isolated Amazonian communities. *Am. J. Epidemiol.* 139: 728-733.

Ganime AC, Leite JP, de Abreu Corrêa A, Melgaço FG, Carvalho-Costa FA, Miagostovich MP. 2015. Evaluation of the swab sampling method to recover viruses from fomites. *J. Virol. Methods* 217: 24-27.

Gallimore CI, Barreiros MA, Brown DW, Nascimento JP, Leite JP. 2004. Noroviruses associated with acute gastroenteritis in a children's day care facility in Rio de Janeiro, Brazil. *Braz. J. Med. Biol. Res.* 37: 321-326.

Garcia LA, Nascimento MA, Barardi CR. 2015. Effect of UV light on the inactivation of recombinant human adenovirus and murine norovirus seeded in seawater in shellfish depuration tanks. *Food Environ. Virol.* 7: 67-75.

Georgiadis S, Pilger DA, Pereira F, Cantarelli VV. 2010. Molecular evaluation of norovírus in patients with acute gastroenteritis. *Rev. Soc. Bras. Med. Trop.* 43: 277-280.

Gouvea V, Santos N, Timenetsky Mdo C, Estes MK. 1994. Identification of Norwalk virus in artificially seeded shellfish and selected foods. *J. Virol. Methods* 48: 177-187.

Hernandez Jd, da Silva LD, Sousa Junior EC, de Lucena MS, Soares L da S, Mascarenhas JD, Gabbay YB. 2016. Analysis of uncommon norovirus recombinants from Manaus, Amazon region, Brazil: GII.P22/GII.5, GII.P7/GII.6 and GII.Pg/GII.1. *Infect. Genet. Evol.* 39: 365371.

Keller R, Justino JF, Cassini ST. 2013. Assessment of water and seafood microbiology quality in a mangrove region in Vitória, Brazil. *J. Water Health* 11: 573-580.

Kojima S, Kaheyama T, Fukushi S, Hoshino FB, Shinohara M, Uchida K, Natori K, Takeda N et al. 2002. Genogroup-specific PCR primers for detection of Norwalk-like viruses. *J. Vir. Methods* 100: 107-114.

Lamounier TA, de Oliveira LM, de Camargo BR, Rodrigues KB, Noronha EF, Ribeiro BM, Nagata T. 2015. Production of Brazilian human norovirus VLPs and comparison of purification methods. *Braz. J. Microbiol.* 46: 1265-1268.

Lindesmith LC, Beltramello M, Donaldson EF, Corti D, Swanstrom J, Debbink K, Lanzavecchia A, Baric RS. 2012. Immunogenetic mechanisms driving norovirus GII.4 antigenic variation. *PLoS Pathog.* 8(5):e1002705.

Luchs A, Morillo SG, Ribeiro CD, Vilanova BC, Calux SJ, Carmona Rde C, Timenetsky M do C. 2011. Gastroenteritis outbreak due to G2P[4] rotavirus and GII norovirus at two correctional facilities in Brazil, 2010. *J. Clin. Virol.* 51: 213-214.

Mahar JE, Kirkwood CD. 2011. Characterization of norovirus strains in Australian children from 2006 to 2008: prevalence of recombinant strains. *J. Med. Virol.* 83: 2213-2219.

Marques Mendanha de Oliveira D, Souza M, Souza Fiaccadori F, César Pereira Santos H, das Dôres de Paula Cardoso D. 2014. Monitoring of Calicivirus among day-care children: evidence of asymptomatic viral excretion and first report of GI.7 Norovirus and GI.3 Sapovirus in Brazil. *J. Med. Virol.* 86: 1569-1575.

Melgaço FG, Victoria M, Corrêa AA, Ganime AC, Malta FC, Brandão ML, de Mello Medeiros V, de Oliveira Rosas C et al. 2016. Virus recovering from strawberries: Evaluation of a skimmed milk organic flocculation method for assessment of microbiological contamination. *Int. J. Food Microbiol.* 217: 14-19.

Melgaço, FG. 2016. Avaliação da metodologia de floculação orgânica para recuperação de vírus entéricos em frutas e queijos [Evaluation of organic flocculation methodology for recovery of enteric virus in fruits and cheeses]. PhD Thesis, Instituto Oswaldo Cruz, Fiocruz.

Menezes FG, Correa VM, Franco FG, Ribeiro MI, Cardoso MF, Morillo SG, Carmona Rde C, Timenetsky M do C et al. 2010. An outbreak of norovirus infection in a long-term care facility in Brazil. *Einstein* (São Paulo) 8: 410-413.

Miagostovich MP, Ferreira FF, Guimarães FR, Fumian TM, Diniz-Mendes L, Luz SL, Silva LA, Leite JP. 2008. Molecular detection and characterization of gastroenteritis viruses occurring naturally in the stream waters of Manaus, central Amazonia, Brazil. *Appl. Environ. Microbiol.* 74: 375-382.

Miagostovich MP, Guimarães FR, Vieira CB, Fumian TM, da Gama NP, Victoria M, de Oliveira JM, Mendes AC et al. 2014. Assessment of water quality in a border region between the Atlantic forest and an urbanised area in Rio de Janeiro, Brazil. *Food Environ. Virol.* 6: 110-115.

Moresco V, Viancelli A, Nascimento MA, Souza DS, Ramos AP, Garcia LA, Simões CM, Barardi CR. 2012. Microbiological and physicochemical analysis of the coastal waters of southern Brazil. *Mar. Pollut. Bull.* 64: 40-48.

Morillo SG, Cilli A, de Cássia Compagnoli Carmona R, Timenetsky M do C. 2008. Identification and molecular characterization of norovirus in São Paulo state, Brazil. *Braz. J. Microbiol.* 39: 619-622.

Morillo SG, Luchs A, Cilli A, Ribeiro CD, Calux SJ, Carmona R de C, Timenetsky M do C. 2011a. Large gastroenteritis outbreak due to norovirus GII in São Paulo, Brazil, summer 2010. *Rev. Inst. Med. Trop. Sao Paulo* 53: 119-120.

Morillo SG, Luchs A, Cilli A, Ribeiro CD, Calux SJ, Carmona Rde C, Timenetsky M do C. 2011b. Norovirus 3rd Generation kit: an improvement for rapid diagnosis of sporadic gastroenteritis cases and valuable for outbreak detection. *J. Virol. Methods* 173: 13-16.

Morillo SG, Luchs A, Cilli A, do Carmo Sampaio Tavares Timenetsky M. 2012. Rapid detection of norovirus in naturally contaminated food: foodborne gastroenteritis outbreak on a cruise ship in Brazil, 2010. *Food Environ. Virol.* 4: 124-129.

Nakagomi T, Correia JB, Nakagomi O, Montenegro FM, Cuevas LE, Cunliffe NA, Hart CA. 2008. Norovirus infection among children with acute gastroenteritis in Recife, Brazil: disease severity is comparable to rotavirus gastroenteritis. *Arch. Virol.* 153: 957-960.

Parks CG, Moe CL, Rhodes D, Lima A, Barrett L, Tseng F, Baric R, Talal A et al. 1999. Genomic diversity of "Norwalk like viruses" (NLVs): pediatric infections in a Brazilian shantytown. *J. Med. Virol.* 58: 426-434.

Prado T, Silva DM, Guilayn WC, Rose TL, Gaspar AM, Miagostovich MP. 2011. Quantification and molecular characterization of enteric viruses detected in effluents from two hospital wastewater treatment plants. *Water Res.* 45: 1287-1297.

Prado T, Guilayn Wde C, Gaspar AM, Miagostovich MP. 2013. The efficiency of concentration methods used to detect enteric viruses in anaerobically digested sludge. *Mem. Inst. Oswaldo Cruz* 108: 77-83.

Prado T, Gaspar AM, Miagostovich MP. 2014. Detection of enteric viruses inactivated sludge by feasible concentration methods. *Braz. J. Microbiol.* 45: 343-349.

Portal TM, Siqueira JA, Costa LC, Lima IC, Lucena MS, Bandeira Rda S, Linhares Ada C, Luz CR et al. 2016. Caliciviruses in hospitalized children, São Luís, Maranhão, 19971999: detection of norovirus GII.12. *Braz. J. Microbiol.* 47: 724-730.

Raboni SM, Damasio GA, Ferreira CE, Pereira LA, Nogueira MB, Vidal LR, Cruz CR, Almeida SM. 2014. Acute gastroenteritis and enteric viruses in hospitalized children in southern Brazil: aetiology, seasonality and clinical outcomes. *Mem. Inst. Oswaldo Cruz* 109: 428-435.

Ribeiro LR, Giuberti RS, Barreira DM, Saick KW, Leite JP, Miagostovich MP, Spano LC. 2008. Hospitalization due to norovirus and genotypes of rotavirus in pediatric patients, state of Espírito Santo. *Mem. Inst. Oswaldo Cruz* 103: 201-206.

Sá AC, Gómez MM, Lima IF, Quetz JS, Havt A, Oriá RB, Lima AA, Leite JP. 2015. Group a rotavirus and norovirus genotypes circulating in the northeastern Brazil in thepostmonovalent vaccination era. *J. Med. Virol.* 87: 1480-1490.

Siebenga JJ, Lemey P, Kosakovsky Pond SL, Rambaut A, Vennema H, Koopmans M. 2010. Phylodynamic reconstruction reveals norovirus GII.4 epidemic expansions and their molecular determinants. *PLoS Pathog.* 6: e1000884.

Silva LD, Rodrigues EL, Lucena MS, Lima IC, Oliveira D de S, Soares LS, Mascarenhas JD, Linhares Ada C et al. 2013. Detection of the pandemic norovirus variant GII.4 Sydney 2012 in Rio Branco, state of Acre, northern Brazil. *Mem. Inst. Oswaldo Cruz* 108: 10681070.

Silva PF, Alfieri AF, Barry AF, de Arruda Leme R, Gardinali NR, van der Poel WH, Alfieri AA. 2015. High frequency of porcine norovirus infection in finisher units of Brazilian pig production systems. *Trop. Anim. Health Prod.* 47: 237-241.

Siqueira JA, Linhares Ada C, Oliveira D de S, Soares L da S, Lucena MS, Wanzeller AL, Mascarenhas JD, Gabbay YB. 2011. Evaluation of third-generation RIDASCREEN enzyme immunoassay for the detection of norovirus antigens in stool samples of hospitalized children in Belém, Pará, Brazil. *Diagn. Microbiol. Infect. Dis.* 71: 391-395.

Siqueira JA, Linhares Ada C, Gonçalves M dos S, Carvalho TC, Justino MC, Mascarenhas JD, Gabbay YB. 2013a. Group A rotavirus and norovirus display sharply distinct seasonal profiles in Belém, northern Brazil. *Mem. Inst. Oswaldo Cruz* 108: 661-664.

Siqueira JA, Linhares Ada C, de Carvalho TC, Aragão GC, Oliveira D de S, Dos Santos MC, de Sousa MS, Justino MC et al. 2013b. Norovirus infection in children admitted to hospital for acute gastroenteritis in Belém, Pará, Northern Brazil. *J. Med. Virol.* 85: 737744.

Siqueira JA, Bandeira RD, Justino MC, Linhares AD, Gabbay YB. 2016. Characterization of novel intragenotype recombination events among norovirus pandemic GII.4 variants. *Infect. Genet. Evol.* 44: 361-366.

Soares CC, Santos N, Beard RS, Albuquerque MC, Maranhão AG, Rocha LN, Ramírez ML, Monroe SS et al. 2007. Norovirus detection and genotyping for children with gastroenteritis, Brazil. *Emerg. Infect. Dis.* 13: 1244-1246.

Souza DS, Ramos AP, Nunes FF, Moresco V, Taniguchi S, Leal DA, Sasaki ST, Bícego MC et al. 2012. Evaluation of tropical water sources and mollusks in southern Brazil using microbiological, biochemical, and chemical parameters. *Ecotoxicol. Environ. Saf.* 76: 153161.

Talal AH, Moe CL, Lima AA, Weigle KA, Barrett L, Bangdiwala SI, Estes MK, Guerrant RL. 2000. Seroprevalence and seroincidence of Norwalk-like virus infection among Brazilian infants and children. *J. Med. Virol*. 61: 117-124.

Teixeira DM, Hernandez JM, Silva LD, Oliveira D de S, Spada PK, Gurjão TC, Mascarenhas JD, Linhares AC et al. 2016. Occurrence of Norovirus GIV in Environmental Water Samples from Belém City, Amazon Region, Brazil. *Food Environ. Virol*. 8:101-104.

Timenetsky M do C, Kisielius JJ, Grisi SJ, Escobar AM, Ueda M, Tanaka H. 1993. Rotavirus, adenovirus, astrovirus, calicivirus and small round virus particles in feces of children with and without acute diarrhea, from 1987 to 1988, in the greater São Paulo. *Rev. Inst. Med. Trop. São Paulo* 35: 275-280.

Vega E, Barclay L, Gregoricus N, Williams K, Lee D, Vinjé J. 2011. Novel surveillance network for norovirus gastroenteritis outbreaks, United States. *Emerg. Infect. Dis*. 17: 13891395.

Vicentini F, Denadai W, Gomes YM, Rose TL, Ferreira MS, Le Moullac-Vaidye B, Le Pendu J, Leite JP et al. 2013. Molecular characterization of noroviruses and HBGA from infected Quilombola children in Espirito Santo State, Brazil. *PLoS One* 8: e69348.

Victoria M, Carvalho-Costa FA, Heinemann MB, Leite JP, Miagostovich M. 2007. Prevalence and molecular epidemiology of noroviruses in hospitalized children with acute gastroenteritis in Rio de Janeiro, Brazil, 2004. *Pediatr. Infect. Dis. J*. 26: 602-606.

Victoria M, Miagostovich MP, Ferreira MS, Vieira CB, Fioretti JM, Leite JP, Colina R, Cristina J. 2009a. Bayesian coalescent inference reveals high evolutionary rates and expansion of Norovirus populations. *Infect. Genet. Evol*. 9: 927-932.

Victoria M, Guimarães F, Fumian T, Ferreira F, Vieira C, Leite JP, Miagostovich M. 2009b. Evaluation of an adsorption-elution method for detection of astrovirus and norovirus in environmental waters. *J. Virol. Methods* 156: 73-76.

Victoria M, Guimarães FR, Fumian TM, Ferreira FF, Vieira CB, Shubo T, Leite JP, Miagostovich MP. 2010a. One year monitoring of norovirus in a sewage treatment plant in Rio de Janeiro, Brazil. *J. Water Health* 8: 158-165.

Victoria M, Rigotto C, Moresco V, de Abreu Corrêa A, Kolesnikovas C, Leite JP, Miagostovich MP, Barardi CR. 2010b. Assessment of norovirus contamination in environmental samples from Florianópolis City, Southern Brazil. *J. Appl. Microbiol.* 109: 231-238.

Victoria M, Fumian TM, Rocha MS, Dalmao F, Leite JP, Girones R, Miagostovich MP. 2014. Gastroenteric virus dissemination and influence of rainfall events in urban beaches in Brazil. *J. Appl. Microbiol.* 117: 1210-1218.

Vieira CB, Mendes AC, Guimarães FR, Fumian TM, Leite JP, Gaspar AM, Miagostovich MP. 2012. Detection of enteric viruses in recreational waters of an urban lagoon in the city of Rio de Janeiro, Brazil. *Mem. Inst. Oswaldo Cruz* 107: 778-784.

Vieira CB, de Abreu Corrêa A, de Jesus MS, Luz SL, Wyn-Jones P, Kay D, Vargha M, Miagostovich MP. 2016. Viruses Surveillance under Different Season Scenarios of the Negro River Basin, Amazonia, Brazil. *Food Environ. Virol.* 8: 57-69.

Vinjé J, Hamidjaja RA, Sobsey MD. 2004. Development and application of a capsid VP1 (region D) based reverse transcription PCR assay for genotyping of genogroup I and II noroviruses. *J. Virol. Methods* 116: 109-117.

White PA. 2014. Evolution of norovirus. *Clin. Microbiol. Infect.* 20: 741-745.

Xavier MP, Oliveira SA, Ferreira MS, Victoria M, Miranda V, Silva MF, Strina A, Barreto ML et al. 2009. Detection of caliciviruses associated with acute infantile gastroenteritis in Salvador, an urban center in Northeast Brazil. *Braz. J. Med. Biol. Res.* 42: 438-444.

In: Noroviruses
Editor: Jesús L. Romalde
ISBN: 978-1-53611-939-8

Chapter 2

NOROVIRUSES IN ITALY

Elisabetta Suffredini[1,*] and Giuseppina La Rosa[2]
[1]Department of Veterinary Public Health and Food Safety
[2]Department of Environment and Primary Prevention,
Istituto Superiore di Sanità, Rome, Italy

ABSTRACT

Noroviruses (NoVs) are the major cause of acute gastroenteritis in children and adults. Virus transmission may occur through contaminated food, water, surfaces or by person-to-person contact. In this chapter we review the available information on the epidemiology of NoVs in the Italian population, and on their prevalence and levels in water environments, food products and food production environments in Italy. A summary of the foodborne and waterborne NoV outbreaks reported in Italy since 2000 is provided.

Keywords: Norovirus, Italy, prevalence, food, water, epidemiology, outbreak

---

* Corresponding Author address: Department of Veterinary Public Health and Food Safety, and Department of Environment and Primary Prevention, Istituto Superiore di Sanità, Rome, Italy. Email: elisabetta.suffredini@iss.it, giuseppina.larosa@iss.it.

EPIDEMIOLOGY OF NOROVIRUS IN ITALY

Acute viral gastroenteritis cases are not subjected to mandatory communication in Italy and no national surveillance system has been established for Noroviruses (NoVs). Therefore, epidemiological data on NoVs circulation in the Italian population are mostly based on hospital-based surveillances and on data, including molecular information on arising strains or variants, provided through collaborative networks as, for example, ISGEV (Italian Study Group for Enteric Viruses; https://isgev.net/).

The surveillance data available in Italy are mostly related to the prevalence of Norovirus acute gastroenteritis in the pediatric population: in a recent study performed in Veneto Region (North-East Italy) on 555 children age <5 years with gastroenteritis symptoms, NoVs accounted for 109 cases (19.6%) with an incidence rate of 1.51 per 100 person-year (Donà et al., 2016). A study on 2834 stool samples collected in the period between 2007 and 2011 from children (age <14 years) with acute gastroenteritis, attending the Hospital of Parma (Northern Italy), detected NoVs in 435 samples (15.3%), largely from children aged <3 years (68.0%) (Medici et al., 2014). Similar incidence of NoV infection was provided in other studies performed in children admitted to hospitals with acute gastroenteritis in Sicily (year 2003, 215 patients, NoV 18.6%; Colomba et al., 2006), Brescia (years 2003-2005, 139 patients, NoV 23.7%; Caracciolo et al., 2007), and Milan (years 2008-2009, 273 patients, NoV 16.2%; Rimoldi et al., 2011). Interestingly, a study conducted on stool samples from pediatric (n = 356) and adult patients (n = 333) with gastrointestinal syndromes, admitted between 2011 and 2012 to the hospital of Pavia (North-East Italy), showed that NoV incidence was <10% in pediatric patients, and increased in the age groups, reaching 51.3% in adults aged >65 years (Rovida et al., 2013).

Besides hospital surveillances, information on the epidemiology of NoVs is occasionally provided by outbreak investigations (see section "Foodborne and waterborne outbreaks of gastroenteritis reported in Italy, associated with Noroviruses" for a review of the available data) and by seroprevalence studies. Among the latter, two studies have been performed in Italy to assess the seroepidemiology of NoV GI, GII and GIV. The first one, performed on 1729 sera collected during 1996 in Verona (North-East Italy) showed the presence of antibodies towards NoV GI (Southampton, GI.2) and NoV GII (Lordsdale, GII.4) respectively in 50.5% and 91.2% of the sera (Pelosi et al., 1999). The second study, that analyzed 535 sera from patients admitted to a hospital in Sicily between 2010 and 2011, showed that antibodies specific for GIV.1 and GIV.2 were present in the tested sera with a prevalence ranging, among the different age groups, from 6.6% to 44.8% for GIV.1 and from 6.8% to 15.1% for GIV.2 (Di Martino et al., 2014).

Occurrence of Norovirus in Different Water Environments in Italy

The presence of human NoVs in sewage and environmental samples impacted by sewage (surface waters, ground water, seawater) has been widely reported in Italy. Moreover, a number of waterborne outbreaks, originated from sewage polluted drinking and recreational waters, have been described. Here we review the state of the art on human NoVs in water environments, and their transmission via the waterborne route in Italy. Table 1 summarizes the studies on NoVs in water environments in Italy.

Raw and Treated Wastewaters

Noroviruses were first described to occur in raw and treated wastewater in 2004-2005 in central Italy (Tuscany region): two of 12 raw sewage samples were positive for NoVs (GI and GII), and 3 out of 12 treated effluents were positive for GII NoVs (Carducci et al., 2006). NoVs were then described in sewage samples in 2006 in two wastewater treatment plants (WTPs) in Rome (La Rosa et al., 2007). Urban sewage samples (N° 10) were 100% positive for NoVs, and 60% presented multiple NoV contaminations (simultaneous presence of both GI and GII). Moreover, mixed infections with different strains of GII NoV were detected in some samples by sequencing of the cloned PCR fragments. Six different genotypes were detected: GI.1, GI.2, GI.4, GII.4, GIIb, and GII.3. Using real-time PCR, sewages were found to contain from 6.8×10^2 to 6.7×10^4 genome copies (GC) per millilitre (La Rosa et al., 2009).

In 2007, 50 grab samples (25 inflows and 25 outflows) collected from five different municipal WTPs in central Italy were tested for NoVs to assess the occurrence and concentration in inflows and outflows, and to obtain a rough estimation of the efficacy of the purification process (La Rosa et al., 2010a). In all, 92% and 68% of the samples were positive by real-time PCR for GII, and 72% and 40% for GI in influents and effluents, respectively. Higher counts for GI as compared to GII were detected in both influent and effluent samples. Removal efficiencies for NoV ranged from 62% (NoV GII) to 78% (NoV GI), indicating that wastewater treatment systems do not completely remove enteric viruses, which can therefore easily reach other water sources.

Another study confirmed the wide circulation of NoVs in urban wastewaters in Italy, with a predominance of GI strains, and the widespread distribution of NoV variants in both raw and treated wastewater (La Rosa et al., 2010b). The study was performed on 97 raw and treated wastewater samples collected in central Italy in 2007, and detected NoVs by nested PCR in 96.9% of influent samples: 89.1% of the samples contained both genogroups, 4.7% contained only GI and 3.1% only GII. In treated effluents, NoVs were detected in 78.8% of samples: 30.3% contained both genogroups, and 48.5% contained only GI. The major genotypes detected by sequencing analyses were GI.2, GI.5, GII.b, GII.4 and GII.6.

Table 1. Detection of NoVs in water environments in Italy

Matrices analyzed	Year	Genogroups and Types identified	Quantity	Main results	Reference
Raw and treated sewage	2004-2005	GI and GII	-	Raw sewage: 2/12 (17%) positive samples; treated sewage: 3/12 (25%) positive samples	Carducci et al., 2006
Raw sewage	2006	GI.1, GI.2, GI.4, GII.4, GIIb, GII.3	From 6.8×10^2 to 6.7×10^4 genome copies (GC) per mL	10/10 (100%) sewage samples positive by nested PCR	La Rosa et al., 2007; La Rosa et al., 2009
Raw and treated sewage	2007	-	Mean GI NoV concentration 9.3×10^6 (influent samples) and 2.0×10^6 (effluent samples); Mean GII NoV concentration: 2.6×10^6 GC/mL (influent samples) and 9.9×10^5 (effluent samples)	GI NoV detected in 92% of raw influents and in 68% of treated effluents; GII NoV detected in 72% of raw influents and in 40% of treated effluents. Removal efficiencies for noroviruses ranged from 62% (GII) to 78% (GI)	La Rosa et al., 2010a
Raw and treated sewage	2007	GI.2, GI.5, GIIb, GII.4 and GII.6	-	88/97 (90.7%) of samples were positive for NoVs (96.9% in influent and 78.8% in effluent samples). Most of the positive samples (69.1%) contained both genogroups, 19.6% contained only GI, while 2.1% only GII	La Rosa et al., 2010b
Raw and treated sewage	2008	-	-	NoVs were detected in 2/27 (7.4%) influent samples, and 5/54 (9.2%) samples collected during treatments; they were not detected in chlorinated effluents	Anastasi et al., 2008; Petrinca et al., 2009
Raw sewage	2009	GI and GII	-	GII was present in raw sewage in 9/9 tested samples (100%) and GI in 7/9 (77.8%)	Carducci et al., 2009
Raw and treated sewage	2013	GI.2, GI.4, and GI.6, GII.1, GII.2, and GII.4	-	NoVs were detected in 10/21 (48%) of raw wastewater samples, 4/12 (19%) of treated effluents, and 1/12 (8%) of ultrafiltered waters. Most of the positive samples (9) contained both GI and GII	Iaconelli et al., 2016

Table 1. (Continued)

Matrices analyzed	Year	Genogroups and Types identified	Quantity	Main results	Reference
Raw sewage	2007	GIV.1	-	GIV NoV detected in 1 out 75 sewage samples, by nested PCR	La Rosa et al., 2008
Raw sewage	2008-2009	GIV.1	-	10/122 (8.2%) sewage samples positive for GIV NoV by nested PCR	La Rosa et al. 2010c
Raw sewage	2011-2012	GIV.1	-	67 /307 (21.8%) sewage samples positive for GIV NoV by nested PCR	Muscillo et al., 2013
Raw and treated sewage	2013	GI and GII	5.83 and 5.80 log_{10} NoV GC/L, detected in influent and effluent wastewaters respectively (mean concentration)	NoV detected in 72% of entry samples and 65% of exit samples.	Carducci and Verani, 2013
River and seawater	2004-2005	G1 and GII	-	4/12 (33%) river waters positive for NoV; 1/12 (8%) seawater samples positive for NoV	Verani et al., 2006; Carducci et al., 2006
River, estuarine, and seawater samples	2004-2005	GII.4, GIIb	7.6×10^1 to 2.4×10^3 GC /10L	River and estuarine samples tested all negative. 3/26 (11.5%) seawater samples positive for GII NoV by nested PCR	La Rosa et al., 2007; La Rosa et al., 2009
Fresh and marine recreational waters	2005-2007	GI and GII	-	GI NoV detected in 37 recreational waters in four sites in Italy. G II NoV detected at marine and freshwater sites, the highest frequencies being 16.3% positive samples at Ardea (Rome, marine)	Wyn-Jones et al., 2011
Lake	2011-2014	GII	10^4 GC/L	NoV detected in lake samples from an agricultural area by means of Real time PCR	D'Ugo et al., 2016
Seawater in shellfish production areas, and water samples from nearby underwater sewage discharge points	2015	GI.1, GI.2, GI.3, GI.4, GI.6, GII.1, GII.2, GII.4 GII.6, and GII.17	-	3/39 (7.7%) seawater samples from shellfish production areas and 10/29 (34.5%) water samples from nearby underwater sewage discharge points tested positive for NoV by nested PCR	Suffredini et al., 2016; la Rosa et al., 2016

GC = Genome copy number

Table 2. Detection of NoVs in food and food production environments in Italy

Matrices analyzed	Year	Genogroups and Types identified	Quantity	Main results	Reference
Bivalve molluscs (primary production)	2003-2004	GII.1, GII.2, GII.4, GIIb	-	235 samples (mussels, clams, oysters, scallops); NoV detected in 36 samples (15.3%)	Croci et al., 2007
Bivalve molluscs (primary production)	2005-2006	GII.4, GIIb	-	120 samples (mussels, clams); NoV detected in 10/120 samples (8.3%)	Suffredini et al., 2008
Bivalve molluscs (primary production)	2008-2009	-	-	70 samples (mussels, clams, oysters); NoV detected in 36 samples (51.4%)	Suffredini et al., 2012
Bivalve molluscs (primary production)	2008-2012	-	Mean NoV concentration: 3.1×10^2 GC/g of DT in class A production areas (range <LOQ-3.3×10^4 GC/g), 1.9×10^3 GC/g of DT in class B production areas (range <LOQ-1.4×10^5 GC/g)	336 samples (mussels, clams, oysters, other species); NoV detected in 173 samples (51.5%); NoV prevalence in class A areas (22.1%) significantly lower than in class B areas (66.3%). Different prevalence among shellfish species (mussels 70.3%, clams 66.0%, oysters 47.8%).	Suffredini et al., 2014
Bivalve molluscs (primary production)	unk	-	-	80 samples (mussels); NoV detected in 1 sample (1.2%)	Serracca et al., 2010
Bivalve molluscs (primary production)	2009-2011	-	-	1266 samples (mussels, clams, oysters, other species); NoV detected in 337 samples (26.6%)	Bazzardi et al., 2014
Bivalve molluscs (primary production)	2011-2012	-	-	59 samples (mussels, razor clams); NoV detected in 9 samples (15.3%)	Fusco et al., 2013
Bivalve molluscs (primary production)	2011-2013	-	-	232 samples (mussels, clams, oysters and other species); NoV detected in 53 samples (22.8%)	La Bella et al., 2014
Bivalve molluscs (retail level and primary production)	unk	-	-	137 samples (mussels, clams, oysters); NoV detected in 3 samples (2.2%); NoV not detected in primary production; prevalence at retail: 4.5%	Gabrieli et al., 2007
Bivalve molluscs (retail level and primary production)	2007-2010	-	-	163 samples (mussels, other species); NoV detected in 94 samples (57.7%). NoV prevalence in production areas: 51.3%. Different prevalence in class A and class B production areas (22.4% vs 79.7%). NoV prevalence at retail: 73.9%	Pepe et al., 2012
Bivalve molluscs (retail level and primary production)	unk	GII.4, GIV.1	-	11 samples (mussels, clams). GI not detected, GII detected in 6 (2 in primary production, 4 at retail level) samples (55%), GIV in 1 sample (9%)	La Rosa et al., 2012

Table 2. (Continued)

Matrices analyzed	Year	Genogroups and Types identified	Quantity	Main results	Reference
Bivalve molluscs (retail level)	2005-2008	GII.4, GIIb	-	116 samples (mussels, clams, oysters); NoV detected in 14 samples (12.1%)	Terio et al., 2010
Bivalve molluscs (retail level)	2015-2016	-	Mean GI concentration: 2.4×10^2 GC/g of DT (range <LOQ-4.6×10^3 GC/g); mean GII concentration: 3.3×10^2 GC/g of DT (range <LOQ-1.7×10^4 GC/g)	89 samples (mussels, clams, oysters and other species). NoV detected in 36 samples (prevalence 40.4%): GI in 19 (21.3%), GII in 32 (35.9%), GIV in 6 (6.7%)	Suffredini et al., 2016
Seafood products (retail level)	2003-2011	GI.3, GII.1, GII.2, GII.4, GII.4_2004, GII.4_2006a, GII.4_2006b, GII.8, GIIb, GIIe	-	4463 samples (mussels, clams, oysters, other shellfish species, preserved products). NoV detected in 183 samples (4.1%). NoV prevalence fluctuated over the years from 0.6% to 9.8%	Pavoni et al., 2013
Raw and ready-to-eat vegetables (retail and production level)	2005-2007	-	-	124 samples; NoV not detected	De Giusti et al., 2010
Ready-to-eat vegetables (retail level)	2010	GII.4	-	110 samples; NoV not detected in leafy vegetables samples (n=80), detected in 12/25 dried tomatoes in bulk (48%) and 3/5 tomatoes in oil (60%)	Serracca et al., 2012
Raw and ready-to-eat vegetables (retail and production level)	2011-2012	-	-	2532 samples; NoV detected in 1/1372 leafy vegetables (0.1%) and not detected in RTE vegetables (n=1160)	Losio et al., 2015
Snails	unk	-	-	28 samples; NoV not detected	Corda et al., 2014
Pig herds	2003-2006	-	-	209 fecal samples; porcine NoV not detected	Martella et al., 2008
Pig herds	2006-2007 and 2012	GII.11 (porcine NoV)	-	290 fecal samples; porcine NoV detected in 1 sample (0.3%)	Di Bartolo et al., 2014
Pig herds	2012-2014	-	-	242 fecal samples; porcine NoV not detected	Monini et al., 2015

GC = Genome copy number; DT = Digestive Tissue

During the period November 2007–April 2008, a total of 18 sewage samples collected in a WTP in Pisa (9 at the entrance and 9 at the exit) were tested for NoV. GII NoVs were present in the raw sewage in all tested samples (100%) and GI NoV in seven of nine (77.8%) (Carducci et al., 2009).

Anastasi and coworkers evaluated the presence of enteric viruses and their removal in three WTPs of different sizes and with different treatment methods, located in the area of Rome. Overall, 92 samples were collected in 2008, during three seasonal campaigns (winter, spring-summer, and autumn): 11 influent samples, 54 samples collected at intermediate treatment points, and 11 chlorinated effluents. NoVs were detected in two influent samples (7.4%), five samples collected during treatments (9.2%), and were not detected in chlorinated effluents (Anastasi et al., 2008; Petrinca et al., 2009).

Quantitative data on NoV in raw and treated wastewaters were obtained on 46 samples (23 entry, 23 exit) collected in February–March 2009 and September 2009–June 2010 from a WTP in Pisa, central Italy. NoVs was detected in 72% of entry samples and 65% of exit samples. The mean concentrations of NoV detected in influent and effluent wastewaters were 5.83 and 5.80 $\log_{10}$ GC/L, respectively (Carducci and Verani, 2013).

More recently, NoVs were detected in 48% (10/21) of raw wastewater samples, 19% (4/12) of treated effluents, and 8% (1/12) of ultrafiltered waters collected in 2013 in Piedmont, northern Italy (Iaconelli et al., 2016). Of the 15 positive samples, nine contained both GI and GII, four had only GII, and two only GI. Three genotypes were detected within GI (GI.2, GI.4, and GI.6), and three within GII (GII.1, GII.2, and GII.4).

The first detection of NoV GIV in water environments in Italy was documented in a raw sewage sample (out of 75 tested) collected in 2007 near Rome, analysed using a newly designed nested RT-PCR assay targeting the capsid region (La Rosa et al., 2008). Using the same PCR assay the authors also detected NoV GIV in three of 19 stool samples collected from paediatric cases with symptoms of acute gastroenteritis in Rome (La Rosa et al., 2008). Phylogenetic analysis showed that the environmental and clinical samples were similar, and grouped with the

GIV.1 human cluster. This was the first report to specifically address the presence of NoV GIV in Italy and to provide molecular characterization of an environmental GIV strain, thus gathering firm evidence for the circulation of this genogroup in Italy. Later on, the same authors conducted, in 2008-2009 and in 2011-2012, two large-scale molecular epidemiological investigations of WTPs throughout Italy and detected GIV in 10/122 samples (8.2%) and 67/307 samples (21.8%) respectively (La Rosa et al. 2010c; Muscillo et al., 2013), confirming that not only GI and GII but also GIV NoV strains are circulating in humans.

River, Estuarine and Seawater Samples

Different authors documented the occurrence of human NoVs in environmental samples impacted by sewage (river, lake, and seawater samples). In the first study conducted in Italy by Verani and coworkers, water samples, collected in years 2004–2005 from a river receiving the effluent of the treatment plant of the city of Pisa, were tested for the presence of NoVs, along with other enteric viruses (Torque teno virus, enteroviruses, and rotaviruses) (Verani et al., 2006). Four out of 12 samples tested positive for NoVs: 8% (1/12) for NoV GI and 25% for NoV GII. In the same years, a monthly monitoring of seawaters detected NoVs in 1/12 (8%) samples, with both GI and GII NoV identified (Carducci et al., 2006). Later on, a study performed on environmental samples from river, estuarine, and seawater samples showed the occurrence of GII NoVs (genotypes GII.4 and GIIb) in 3 out of 26 seawater samples (11.5%) (La Rosa et al., 2007), while river and estuarine samples tested all negative for NoVs. Seawater samples were found to contain from 7.6×10^1 to 2.4×10^3 genome copies (GC) per 10 L (La Rosa et al., 2009).

Data on the occurrence of NoVs in fresh and marine recreational waters in Italy were obtained in the period 2005-2007 in the framework of a large European project called VIROBATHE (Methods for the Detection of Adenoviruses and Noroviruses in European Bathing Waters with Reference to the Revision of the Bathing Water Directive 76/160/EEC) conducted to acquire data about NoVs and Adenoviruses, and to standardize reliable techniques for the detection and quantification of such viruses in water samples for water quality assessment. NoVs were present in fresh and marine recreational waters in Europe and were detectable with varying frequency. Almost all the NoV GI-positive samples (37/38) were found in four sites in Italy, with a marine site at Pomezia (Rome) showing 30% of positive samples. Genogroup II NoVs were detected at marine and freshwater sites, the highest frequencies being 16.3% positive samples at Ardea (Rome, marine). However, in the areas in Italy close to the points of collection of the NoV-positive environmental samples, there was no evidence of norovirus-related outbreaks at the time when the samples were taken (Wyn-Jones et al., 2011).

Another European project μAQUA (Universal microarrays for the evaluation of fresh-water quality based on detection of pathogens and their toxins) studied the presence NoVs (along with other enteric viruses) in freshwater samples in Italy and documented the detection, by real tiem-PCR, of GII NoVs in lake samples (10^4 GC/L) from an agricultural area (D'Ugo et al., 2016).

More recently (2015), NoVs has been detected in central Italy in shellfish in production areas, in 3/39 (7.7%) seawater samples and in 10/29 (34.5%) water samples from nearby underwater sewage discharge points, with a number of genotypes detected: GI.1, GI.2, GI.3, GI.4, GI.6, GII.1, GII.2, GII.4 GII.6, and GII.17 (Suffredini et al., 2016; La Rosa et al., 2016).

Groundwater Samples

Only one study is available on NoVs in groundwater samples, and documented the presence of NoV GI and GII in 15.38% (4 out of 26) of samples collected in 2009 in Region Lazio, central Italy (Gabrieli et al., 2009).

Occurrence of Norovirus in Food and Food Production Environments in Italy

The presence of NoVs in bivalve molluscs and the role of this food category in NoV transmission is well documented in literature and several surveys have been undertaken in Italy to assess the prevalence, the quantitative levels and the seasonal trends of viral contamination in shellfish, both in production areas (primary production) and at retail level. Furthermore, in light of the increased awareness of the role of food of non-animal origin in the transmission of foodborne diseases, studies have been conducted also to assess the presence of NoVs in vegetables. Table 2 summarizes the studies on NoVs presence in different foods and food production environments in Italy.

Bivalve Shellfish

Given the high economic and trade relevance of shellfish production in Italy, NoV presence in bivalve shellfish at the production sites and at retail level has been the object of several studies over the years.

In a monitoring program performed in 2003-2004 in the Northern Adriatic Sea, one of the most relevant Italian shellfish production areas with regards to clams, NoVs were detected (occasionally in association with other viruses as hepatitis A), in 15.3% of the 235 tested samples, with contamination ranging from 0% of oysters to 28% of clam samples (Croci et al., 2007). The NoV strains detected in the study included mostly GII genotypes, with a predominance of GII.1, GII.2 and GIIb, and a limited number of samples showing the presence of GII.4 strains. A subsequent study performed in a limited area of the same northern Adriatic region, the delta of the Po river, subjected to the influence of the river on the environmental (salinity, pH, nutrients supply, etc.) and microbial characteristics of the production areas, performed the analysis of 120 bivalve molluscs samples and detected the presence of NoVs in 8.3% of the samples (Suffredini et al., 2008). In this study, NoV contamination was more frequently detected in samples collected from the lagoon areas than from the off-shore production areas included in the study, and the detected strains belonged prevalently to the GII.4 genotype, with a shift that may be a reflection of the appearance of the new GII.4 'Hunter' variant in 2004. The following assessments of the prevalence of NoVs in shellfish production areas of the delta of the Po river provided evidence of a higher circulation of the virus. The analysis of 70 shellfish samples (including approximately an equal number of mussels, clams and oysters) with a standardized real-time RT-PCR protocol, in fact, detected the presence of NoVs in 36 of the 70 samples (51.4%). The prevalence data for the two genogroups showed a predominance of GII respect to GI (48.6% *vs* 37.1%) as well as the tendency to have, in bivalve samples, the co-presence of both genogroups (34.3%) (Suffredini et al., 2012).

With regard to the quantitative levels of NoVs in bivalve molluscs, a study by Suffredini et al. (2014) reassessing viral contamination in 336 samples collected during several monitoring programs on geographically diverse Italian production areas (Veneto, Sardinia, Lazio and Campania Regions) estimated that both NoV prevalence and levels varied in a statistically significant way in relation to the microbiological quality of the growing areas (class A or B according to European Regulation 854/2004). The prevalence of NoVs estimated in this study was, in fact, 22.1% for class A and 66.3% for class B production areas with an average contamination of 3.1×10^2 genome copies (GC) per gram in the first group, and 4.9×10^3 GC/g in the latter. In this study also the contamination ranges of the positive samples varied in the two groups (highest NoV concentrations detected: 3.3×10^4 GC/g in class A areas, 1.4×10^5 GC/g in class B areas) and a significant variability of NoV prevalence was shown depending on the geographic region of origin of the samples.

The geographic variability in NoV prevalence data in Italy is indeed evident by comparison of recent studies performed in bivalve production areas of several Regions. A study conducted on 62 samples collected from shellfish production areas in Region Lazio (Central Thyrrenian Sea) in the period 2005-2008 did not detect NoVs in any of the samples (Gabrieli et al., 2007) and prevalence of NoVs in 80 samples from the Gulf of La Spezia (northern Thyrrenian Sea) was also low (1.2%) (Serracca et al., 2010). On the contrary, a large, multiannual monitoring plan undertaken in one of the most relevant national production areas, Sardinia (Bazzardi et al., 2014), analysed a total of 1266 samples (1014 mussels, 148 clams, 53 oysters and other species) collected from different production areas of the Region and detected the presence of NoVs in 337 samples (26.6%). Similarly, the prevalence of NoVs in bivalve shellfish produced in Region Campania (southern Thyrrenian Sea) varied from 15.3% (Fusco et al., 2013) to 51.3% (Pepe et al., 2012) depending on the study, and analysis of bivalve molluscs collected in one production area in Sicily showed the presence of NoVs in 2 of 2 samples (La Rosa et al., 2012). Finally, data from the Adriatic Sea showed a prevalence ranging from 22.8% of Apulia Region, southern Adriatic Sea (La Bella et al., 2014) to 51.4% of the delta of the Po river, northern Adriatic Sea (Suffredini et al., 2014).

Compared to the data available on primary production of bivalve molluscs, the information on prevalence and levels of NoVs in shellfish at retail level in Italy are relatively limited. Three studies, performed in Region Lazio, Region Apulia and Region Sicily with the use of conventional RT-(semi)nested-PCR protocols, showed the presence of NoVs respectively in 14 of 116 (12.1%) samples (Terio et al., 2010), 3 of 67 (4.5%) samples (Gabrieli et al., 2007) and 6 of 11 (55%) samples (La Rosa et al., 2012). Furthermore, the latter study provided the first evidence of NoV GIV presence in shellfish in Italy (1 of 11 samples, 9%). Another study performed in Region Campania by means of real-time RT-PCR between 2007 and 2010, indicated a much higher NoV prevalence, 73.9% (Pepe et al., 2012). These data have been partially confirmed by a sampling program established in 2015-2016 in markets, retail shops, supermarkets and wholesalers in Naples metropolitan areas, in which NoV prevalence in bivalve molluscs at retail level exceed 40% (Suffredini et al., 2016). In this study, the average contamination levels in the marketed samples was 2.4×10^2 GC/g for GI (range <LOQ-4.6×10^3 GC/g) and 3.3×10^2 GC/g for GII (range <LOQ-1.7×10^4 GC/g) and presence of NoV GIV in retail samples marketed in Italy was confirmed in 6.7% of samples. The high variability

of the results assessing the presence of NoV in bivalve molluscs at retail level may be partially explained with the different analytical methods applied in the studies, with the number of samples analyzed or with the limited geographical and temporal breadth of the monitoring plans. A large retrospective study on the results of analyses conducted between 2003 and 2011 on 4463 seafood samples (mainly mussels and clams) collected in northern Italy showed an average NoV prevalence of 4.1% in marketed products, with values ranging from 1.9% of preserved food to 4.7% of mussels and from 0.6% in 2007 to 9.8% of 2003 (Pavoni et al., 2013).

Raw and Ready-to-Eat Vegetables

The presence of NoVs in food matrices different than bivalve molluscs has been investigated in a limited number of studies in Italy, though the increased awareness of the risk associated to consumption of food of non-animal origin has boosted the development of monitoring plans for the assessment of the prevalence of enteric viruses in vegetables.

In the first study performed in Italy on this food category, 124 whole and ready-to-eat leafy vegetables collected in central Italy were analyzed and NoVs were never detected (De Giusti et al., 2010). A subsequent work by Serracca and colleagues (2012) on ready-to-eat leafy vegetables (80 samples) and dried tomatoes (25 samples in bulk, 5 samples of tomatoes in oil) collected in 2010 in markets and supermarkets in the city of La Spezia (northern Italy) highlighted the high prevalence of contamination in dried tomatoes (48% in tomatoes in bulk, 60% in tomatoes in oil), irrespectively of the origin (imported or national production); in all such samples, the NoV strain detected showed high similarity to the GII.4 genotype. On the contrary, contamination with NoVs of ready-to-eat leafy vegetables was never detected (Serracca et al., 2012). Interestingly, among the viruses included in this study (hepatitis A, hepatitis E, Adenovirus and Norovirus), NoVs were the only viral contaminants detected in the samples. Finally, a large collaborative study performed on 2532 raw and ready-to-eat leafy vegetables collected between 2011 and 2012 in several Italian Regions detected the presence of NoVs in only one sample, with an overall prevalence of contamination of 0.1% (Losio et al., 2015).

Other Foods and Animals for Food Production

Very few studies have attempted assessing the relevance of food matrices, particularly those consumed in traditional and regional dishes, as a potential vehicle of transmission of NoVs. In Italy, one study of this kind has evaluated the prevalence of viral contamination in snails (*Helix aspersa* and *Helix aperta*) produced in local snail farms or imported. A total of 28 samples were tested in the study and none of the samples showed the presence of NoVs (Corda et al., 2014).

Given the important role of pig husbandry in Italian food production, other studies have been performed to evaluate the significance of such animals as a possible reservoir for NoVs and the potential for zoonotic transmission of NoVs. Three studies have been performed to this aim: in the first study (Martella et al., 2008), that analyzed 209 fecal samples collected from diarrheic piglets in Central and northern Italy, NoVs were never detected. The second study (Di Bartolo et al., 2014) analyzed 290 fecal samples collected from asymptomatic and diarrheic pigs in several farms (years 2006-2007 and 2012) and detected the presence of a GII.11 strain (porcine NoV) in one sample (prevalence 0.3%). Finally, the third study (Monini et al., 2015) analyzed 242 pig fecal samples collected in the period 2012-2014 from eight herds located in different parts of the national territory. In this study no NoV-positive sample was detected, therefore confirming the results of the previous studies on the scarce diffusion of NoVs in pigs in Italy.

Foodborne and Waterborne Outbreaks of Gastroenteritis Reported in Italy, Associated with Norovirus

From 2000 to 2013 thirteen NoV outbreaks have been described in Italy, in association with consumption of contaminated drinking water or contaminated food or in relation to exposure to contaminated surface waters. Table 3 summarizes the studies on foodborne and waterborne Norovirus outbreaks in Italy. Overall, waterborne outbreaks accounted for seven of the reported episodes and 4275 cases, and foodborne outbreaks, usually of smaller size, accounted for six of the episodes and 582 cases.

Waterborne Outbreaks in Italy

With regards to waterborne transmission, the first confirmed outbreak of NoV infection in Italy occurred in 2000 at a tourist resort in the Gulf of Taranto (Boccia et al., 2002) in southern Italy. A total of 344 people were involved, 69 of whom were staff members. NoVs were found in 22 of 28 stool specimens tested. The source of illness was likely contaminated drinking water, as environmental inspection identified a breakdown in the resort water system and tap water samples were contaminated with fecal bacteria. Attack rates were increased (51.4%) in staff members involved in water sports. Relative risks were significant only for exposure to beach showers and consuming drinks with ice.

In the summer of 2003, a gastroenteritis outbreak with 183 cases (169 guests and 14 resort employees) due to Norovirus, Rotavirus, and *Campylobacter*, occurred at a holiday resort in central Italy and was associated with contaminated water (Migliorati et al., 2008a; Migliorati et al., 2008b). The case-control study identified sea bathing, use of cabin and shared toilets, and common shower (supplied with non–drinking water) as statistically significant risk factors. Stool samples taken from people with gastroenteritis were found positive for Norovirus (68%), Rotavirus (38%)

and *Campylobacter* (7%). Environmental investigations revealed fecal contamination of the groundwater and the presence of NoVs in the seawater near the resort. Fecally contaminated groundwater and seawater were leaking into the non-drinking-water system, which was found to be connected to the drinking-water system of the resort. The groundwater involved in contamination of both drinking-water and non–drinking-water systems was probably polluted by the river flowing into the sea nearby and by seawater infiltration.

In 2005 a large outbreak of NoV gastroenteritis occurred in the southern Italy region of Puglia (Rizzo et al., 2007), involving 400 guests in a resort in the province of Lecce. Ninety percent of total stool samples examined were positive by RT-PCR, and sequence analysis confirmed the presence of a GII NoV. An epidemiological investigation was performed in order to assess risk factors associated with illness. The most likely hypothesis was that illness was associated with raw mussels and ice made with tap water. This hypothesis could not be confirmed by testing food or ice for NoV contamination. Although laboratory testing for NoVs could not be performed on drinking water, the presence of fecal bacteria suggested that the water system may have been the source of NoVs.

In 2006, an outbreak of gastroenteritis involving 2860 cases in Apulia due to Norovirus and Rotavirus was caused by contaminated water (Martinelli et al., 2007). The epidemiological investigation and the laboratory tests showed that the possible source of infection was the drinkable tap water contaminated with (at least) rotaviruses and NoVs. Microbiological investigations of the pipelines and wells of the water distribution system did not reveal the source of contamination even though technical problems at the local chlorination treatment facilities could not be excluded.

In 2009, an outbreak of gastroenteritis was reported in Lombardy, northern Italy. Over one month 299 persons fulfilled the outbreak case definition. Stool samples tested positive for at least one of the following viral groups: Norovirus, Rotavirus, Enterovirus or Astrovirus. Water samples and filters from the water system also tested positive for Norovirus and Enterovirus (Scarcella et al., 2009). Characterization of viruses from 40 patient stool samples and 5 environmental samples identified three distinct Rotavirus and five Norovirus genotypes; two of the latter were detected in both patient and environmental samples (Di Bartolo et al., 2011).

Table 3. Foodborne and waterborne NoV outbreaks reported in Italy

Year	N° of cases involved	Source	Detection of NoV in food/ water samples	Suspected or identified source	Reference
2000	344	waterborne	No	Drinking water	Boccia et al., 2002
2002	103	foodborne	Yes	Raw shellfish	Prato et al., 2004
2002	202	foodborne	Yes	Oysters	Le Guyader et al., 2006
2003	183	waterborne	Yes	Non–drinking water (sea bathing, shower)	Migliorati et al., 2008a; Migliorati et al., 2008b
2005	400	waterborne	No	Ice from tap water	Rizzo et al., 2007
2005	184	foodborne	No	Seafood salad	Gallo et al., 2007; Di Bartolo et al., 2007
2006	2860	waterborne	No	Tap water	Martinelli et al., 2007
2007	65	foodborne	Yes	Oysters	Di Bartolo et al., 2007
2007	25	foodborne	No	Raw shellfish	D'Agaro et al., 2008
2009	299	waterborne	Yes	Municipal drinking water	Scarcella et al., 2009; Di Bartolo et al., 2011
2011	33	waterborne	Yes	River water	Di Bartolo et al., 2015
2011	156	waterborne	No	Municipal drinking water	Purpari et al., 2012; Giammanco et al., 2014
2013	3	foodborne	Yes	Oysters	Suffredini et al., 2013

In September 2011, an acute gastroenteritis outbreak affected 33 guests of a hotel in northern Italy who developed acute symptoms after a daily excursion in the countryside near a small river (Di Bartolo et al., 2015). All patients had drunk river water during an excursion. Identical GI.4 norovirus genomes were detected from one patient's stools and from the river water. Improper discharge of human sewage into the river may have caused this waterborne outbreak. No risk factor was identified other than participation in the river-side excursion. In particular, no correlation was found with consumption of any food.

In 2011 an outbreak involving 156 cases of NoV infection in Agrigento, Sicily, was ascribed to contaminated water (Purpari et al., 2012; Giammanco et al., 2014). Stool specimens were collected from 12 cases and genotype GII.4 variant New Orleans 2009 was identified in stool samples from 11 of 12 cases tested (91.7%). Epidemiological investigations suggested a possible association with municipal drinking water consumption. Information collected through the questionnaire revealed that 94% of them regularly drank the water from the municipal supply system, while no association was found with exposure to any specific food. Water samples from the public water system were tested for NoVs and a variety of genotypes were detected during the first 3 months of surveillance, including GII.4 strains belonging to different variants from that involved in the gastroenteritis outbreak. However, a conclusive microbiological evidence for a waterborne NoV epidemic was lacking due to the failure to identify an identical NoV genome sequence in both patients and water.

Foodborne Outbreaks in Italy

The first foodborne outbreak reported in Italy occurred in 2002 in the province of Bari, Apulia Region, South-East Italy, during the Easter holidays (March 31st – April 7th). A total of 103 cases were detected by the Regional Epidemiological Office, with 22 probably secondary cases. The retrospective cohort study showed that eating raw shellfish was the principal risk factor for the disease (risk ratio: 1.50) and for mussels consumption an attack rate of 86.3% was observed. Laboratory analysis confirmed the presence of NoVs in 24 of 24 clinical specimens from the cases and in 5 of 11 shellfish samples. Sequencing showed heterogeneity of strains, with simultaneous occurrence of GI and GII strains. In 19 cases and 3 mussels samples an identical GII sequence was detected, while 3 cases and 2 mussel samples formed two separate clusters with higher identity with another GII reference strain; finally 2 cases and one mussel sample matched a GI sequence (Prato et al., 2004).

In December of the same year an oyster-related outbreak occurred in the city of La Spezia (northern Italy) and, simultaneously, in several areas in Paris and in South-East France. Clinical samples and oyster samples (leftovers from the consumed meals and samples from the production area in France) were collected and analyzed. In all, 202 cases were reported in Italy and 127 cases in France. The six stool samples typed in France showed the presence of several genotypes (GII.4 in three cases, GIIb, GI.4 and the simultaneous presence of GII.4 and GI.4 in the other cases). The twenty NoV-positive stool samples typed in Italy contained GII.4 (n = 7), GII.8 (n = 9), GI.6 (n = 3) and GI.4 (n = 1). From the four oysters that were found positive for NoV and that could be typed, a similar variety of genotypes could be detected (one GII.4, one GII.8, and two GI.4 sequences), matching the cases to the suspected food (Le Guyader et al., 2006). This was the first case of international NoV outbreak involving Italy and highlighted the relevance of transboundary networking for the investigation and control of NoV-associated episodes.

In 2005 and 2007, three separate outbreaks took place in North-East Italy (Region Friuli Venezia Giulia) in association to consumption of seafood. The first one, that took place in December 2005 in the province of Udine, involved 184 cases that had participated to Christmas lunches and dinners in a restaurant. The epidemiological study showed association of the cases to the consumption of seafood salad (risk ratio 5.1). The laboratory analysis showed identity of the NoV strains in 14 of the cases (GII.6) and detection of NoVs in an environmental sample (swab) taken from the kitchen surface used for the preparation of seafood appetizers; NoVs, however, were not detected in any food sample taken from the restaurant. The investigation ascertained that 6 of the 15 people working in the restaurant had experienced gastroenteritis symptoms before the outbreak and that a kid in the family of the restaurant's owner had probably been the index case (Gallo et al., 2007; Di Bartolo et al., 2007). The second outbreak took place in January 2007 and involved 65 cases in three separate restaurants, following consumption of oysters from the same batch. In this outbreak episode, the analysis of fecal samples and of oyster residues showed the simultaneous presence of more than one NoV strain in the samples, with a match between the sequences obtained from the cases and the food (Di Bartolo et al., 2007). In the last outbreak, reported in March 2007 in a group of adults who had attended a restaurant of Umago (Croatia), 25 people were involved (attack rate 78%) and the epidemiological investigation suggested a link with a meal of "fasolari" shellfish eaten uncooked (risk ratio: 6.7). According to the sequencing data, the genotype involved in this outbreak was the GII.4 (D'Agaro et al., 2008).

The last foodborne outbreak reported in literature, involved three persons in Massa Carrara (northern Tuscany) who had eaten oysters in a restaurant in the city of La Spezia. The meal had taken place on 12.01.2013 and cases were reported to the National Health System on 14.01.2013; cases were attributed to NoV on the basis of symptomatology and disease onset, as clinical specimens were not available for testing. Investigation at the restaurant led to the collection of two oyster samples: analysis of the leftovers of the oysters consumed in the meal showed contamination levels of 2.1×10^3 GC/g for GI and 3.1×10^3 GC/g for GII, while another batch of oysters present in the restaurant and harvested from the same production area of the oysters suspected for the outbreak, had contamination levels of 2.2×10^3 GC/g for NoVGI and 3.0×10^3 GC/g for NoVGII (Suffredini et al., 2013).

References

Anastasi P, Bonanni E, Cecchini G, Divizia M, Donia D, Di Gianfilippo F, Gabrieli R, Petrinca AR, Zanobini A. 2008. Virus removal in conventional wastewater treatment process. *Ig Sanita Pubbl.* 64: 313-330.

Bazzardi R, Fattaccio MC, Salza S, Canu A, Marongiu E, Pisanu M. 2014. Preliminary study on Norovirus, Hepatitis A Virus, *Escherichia coli* and their potential seasonality in shellfish from different growing and harvesting areas in Sardinia region. *Ital J Food Saf* 3: 1601.

Boccia D, Tozzi AE, Cotter B, Rizzo C, Russo T, Buttinelli G, Caprioli A, Marziano ML, Ruggeri FM. 2002. Waterborne outbreak of Norwalk-like virus gastroenteritis at a tourist resort, Italy. *Emerg Infect Dis* 8: 563–568.

Caracciolo S, Minini C, Colombrita D, Foresti I, Avolio M, Tosti G, Fiorentini S, Caruso A. 2007. Detection of sporadic cases of Norovirus infection in hospitalized children in Italy. New Microbiol 30: 49-52.

Carducci A, Verani M. 2013. Effects of bacterial, chemical, physical and meteorological variables on virus removal by a wastewater treatment plant. *Food Environ Virol* 5: 69–76.

Carducci, A, Battistini, R, Rovini, E and Verani, M. 2009. Viral removal by wastewater treatment: monitoring of indicators and pathogens. *Food EnvironVirol* 1: 85–91.

Carducci A, Verani M, Battistini R, Pizzi F, Rovini E, Andreoli E, Casini B. 2006. Epidemiological surveillance of human enteric viruses by monitoring of differentenvironmental matrices. *Water Sci Technol* 54: 239–244.

Colomba C, De Grazia S, Giammanco GM, Saporito L, Scarlata F, Titone L, Arista S. 2006. Viral gastroenteritis in children hospitalised in Sicily, Italy. *Eur J Clin Microbiol Infect Dis* 25: 570-575.

Corda A, Mara L, Virgilio S, Pisanu M, Chessa G, Parisi A, Cogoni MP. 2014. Microbiological and chemical evaluation of *Helix* spp. snails from local and non-EU markets, utilised as food in Sardinia. *Ital J Food Saf* 3: 1732.

Croci L, Losio MN, Suffredini E, Pavoni E, Di Pasquale S, Fallacara F, Arcangeli G. 2007. Assessment of human enteric viruses in shellfish from the northern Adriatic sea. *Int J Food Microbiol* 114: 252-257.

D'Agaro P, Gallo T, Savoia A, Santon D, Busetti M, Di Bartolo I, Ruggeri FM, Campello C. 2008. Outbreak of Norovirus gastroenteritis in group of adults in North-East Italy. In: ESCAIDE (European Scientific Conference on Applied Infectious Disease Epidemiology) 2008; *Abstract book*; p.244. Berlin, 19-21 November 2008.

De Giusti M, Aurigemma C, Marinelli L, Tufi D, De Medici D, Di Pasquale S, De Vito C, Boccia A. 2010. The evaluation of the microbial safety of fresh ready-to-eat vegetables produced by different technologies in Italy. *J Appl Microbiol* 109: 996-1006.

Di Bartolo I, Gallo T, Zuliani M, Gauci C, Ruggeri FM. 2007. Epidemie di gastroenterite acuta causate da Norovirus trasmessi per via alimentare. In: II Workshop nazionale di virologia veterinaria (2007), *Abstract book*; ISTISAN Congressi 07/C3; p.28.

Di Bartolo I, Monini M, Losio MN, Pavoni E, Lavazza A, Ruggeri FM. 2011. Molecular characterization of noroviruses and rotaviruses involved in a large outbreak of gastroenteritis in Northern Italy. *Appl Environ Microbiol* 77: 5545-5548.

Di Bartolo I, Pavoni E, Tofani S, Consoli M, Galuppini E, Losio MN, Ruggeri FM, Varisco G. 2015. Waterborne norovirus outbreak during a summer excursion in Northern Italy. *New Microbiologica* 38: 109-112.

Di Bartolo I, Tofani S, Angeloni G, Ponterio E, Ostanello F, Ruggeri FM. 2014. Detection and characterization of porcine caliciviruses in Italy. *Arch Virol* 159: 2479-2484.

Di Martino B, Di Profio F, Ceci C, Di Felice E, Green KY, Bok K, De Grazia S, Giammanco GM, Massirio I, Lorusso E, Buonavoglia C, Marsilio F, Martella V. 2014. Seroprevalence of norovirus genogroup IV antibodies among humans, Italy, 2010-2011. *Emerg Infect Dis*. 20: 1828-1832.

Donà D, Mozzo E, Scamarcia A, Picelli G, Villa M, Cantarutti L, Giaquinto C. 2016. Community-acquired Rotavirus gastroenteritis compared with Adenovirus and Norovirus gastroenteritis in Italian children: A Pedianet study. *Int J Pediatr* 2016: 236-243.

D'Ugo E, Marcheggiani S, Fioramonti I, Giuseppetti R, Spurio R, Helmi K, Guillebault D, Medlin LK, Simeonovski I, Boots B, Breitenbach U, Koker L, Albay M, Mancini L. 2016. Detection of human enteric viruses in freshwater from European countries. *Food Environ Virol* 8: 206-214.

Fusco G, Aprea G, Galiero G, Guarino A, Viscardi M. 2013. *Escherichia coli*, *Salmonella* spp., Hepatitis A virus and norovirus in bivalve molluscs in Southern Italy. *Vet Ital* 49: 55-58.

Gabrieli R, Macaluso A, Lanni L, Saccares S, Di Giamberardino F, Cencioni B, Petrinca AR, Divizia M. 2009. Enteric viruses in molluscan shellfish. *New Microbiol* 30: 471-475.

Gabrieli R, Maccari F, Ruta A, Panà A, Divizia M. 2009. Norovirus Detection in groundwater. *Food Environ Virol* 1: 92–96.

Gallo T, Di Bartolo I, Binkin N, Ruggeri FM, Zuliani M. 2005. Un episodio di tossinfezione da Norovirus in un ristorante in provincia di Udine, Dicembre 2005. In: VI Workshop Nazionale Enter-net Italia (2007). *Abstract book;* ISTISAN Congressi 07/C1; p.46.

Giammanco GM, Di Bartolo I, Purpari G, Costantino C, Rotolo V, Spoto V, Geraci G, Bosco G, Petralia A, Guercio A, Macaluso G, Calamusa G, De Grazia S, Ruggeri FM, Vitale F, Maida CM, Mammina C. 2014. Investigation and control of a Norovirus outbreak of probable waterborne transmission through a municipal groundwater system. *J Water Health* 12: 452-464.

Iaconelli M, Muscillo M, Della Libera S, Fratini M, Meucci L, De Ceglia M, Giacosa D, La Rosa G. 2016. One-year surveillance of human enteric viruses in raw and treated wastewaters, downstream river waters, and drinking waters. *Food Environ Virol* 28. doi:10.1007/s12560-016-9263-3.

La Bella G, D'Alessandro M, La Salandra G. 2014. Monitoraggio delle contaminazioni da Norovirus e virus dell'Epatite A in molluschi bivalve in Puglia. In: V Workshop Nazionale di Virologia Veterinaria (2014). *Abstract book*; ISTISAN Congressi 14/C3; p.57.

La Rosa G, Della Libera S, Iaconelli M, Proroga YTR, De Medici D, Martella V, Suffredini E. 2017. Detection of Norovirus GII.17 Kawasaki 2014 in shellfish, marine water and underwater sewage discharges in Italy. *Food Environ Virol* (submitted manuscript).

La Rosa G, Fontana S, Di Grazia A, Iaconelli M, Pourshaban M, Muscillo M. 2007. Molecular identification and genetic analysis of Norovirus genogroups I and II in water environments: comparative analysis of different reverse transcription-PCR assays. *Appl Environ Microbiol* 73: 4152-4161.

La Rosa G, Fratini M, Spuri Vennarucci V, Guercio A, Purpari G, Muscillo M. 2012. GIV noroviruses and other enteric viruses in bivalves: a preliminary study. *New Microbiol* 35: 27-34.

La Rosa G, Iaconelli M, Pourshaban M, Fratini M, Muscillo M. 2010c. Molecular detection and genetic diversity of norovirus genogroup IV: a yearlong monitoring of sewage throughout Italy. *Arch Virol.* 155: 589-593.

La Rosa G, Iaconelli M, Pourshaban M, Muscillo M. 2010b. Detection and molecular characterization of noroviruses from five sewage treatment plants in central Italy. *Water Res* 44: 1777-1784.

La Rosa G, Pourshaban M, Iaconelli M, Muscillo M. 2008. Detection of genogroup IV noroviruses in environmental and clinical samples and partial sequencing through rapid amplification of cDNA ends. *Arch Virol.* 153: 2077-2083.

La Rosa G, Pourshaban M, Iaconelli M, Muscillo M. 2010a. Quantitative real-time PCR of enteric viruses in influent and effluent samples from wastewater treatment plants in Italy. *Ann Ist Super Sanita* 46: 266-73.

La Rosa G, Pourshaban M, Iaconelli M, Muscillo, M. 2009. Quantification of Norovirus genogroups I and II in environmental and clinical samples using TaqMan Real-Time RT-PCR. *Food Environ Virol* 1: 15–22.

Le Guyader FS, Bon F, DeMedici D, Parnaudeau S, Bertone A, Crudeli S, Doyle A, Zidane M, Suffredini E, Kohli E, Maddalo F, Monini M, Gallay A, Pommepuy M, Pothier P, Ruggeri FM. 2006. Detection of multiple noroviruses associated with an international gastroenteritis outbreak linked to oyster consumption. *J Clin Microbiol* 44: 3878-3882.

Losio MN, Pavoni E, Bilei S, Bertasi B, Bove D, Capuano F, Farneti S, Blasi G, Comin D, Cardamone C, Decastelli L, Delibato E, De Santis P, Di Pasquale S, Gattuso A, Goffredo E, Fadda A, Pisanu M, De Medici D. 2015. Microbiological survey of raw and ready-to-eat leafy green vegetables marketed in Italy. *Int J Food Microbiol*. 210: 88-91.

Martella V, Bányai K, Lorusso E, Bellacicco AL, Decaro N, Mari V, Saif L, Costantini V, De Grazia S, Pezzotti G, Lavazza A, Buonavoglia C. 2008. Genetic heterogeneity of porcine enteric caliciviruses identified from diarrhoeic piglets. *Virus Genes* 36: 365-373.

Martinelli D, Prato R, Chironna M, Sallustio A, Caputi G, Conversano M, Ciofi Degli Atti M, D'Ancona FP, Germinario CA, Quarto M. 2007. Large outbreak of viral gastroenteritis caused by contaminated drinking water in Apulia, Italy, May - October 2006 *Euro Surveill* 12: e070419.1.

Medici MC, Tummolo F, Martella V, Chezzi C, Arcangeletti MC, De Conto F, Calderaro A. 2014. Epidemiological and molecular features of norovirus infections in Italian children affected with acutegastroenteritis. *Epidemiol Infect* 142: 2326-2335.

Migliorati G, Prencipe V, Ripani A, Di Francesco C, Casaccia C, Crudeli S, Ferri N, Giovannini A, Marconi MM, Marfoglia C, Melai V, Savini G, Scortichini G, Semprini P, Ruggeri FM. 2008. Gastroenteritis outbreak at holiday resort, Central Italy. *Emerg Infect Dis* 14: 474-478.

Migliorati G, Prencipe V, Ripani A, Di Francesco C, Casaccia C, Crudeli S, Ferri N, Giovannini A, Marconi MM, Marfoglia C, Melai V, Savini G, Scortichini G, Semprini P, Ruggeri FM. 2008. An outbreak of gastroenteritis in a holiday resort in Italy: epidemiological survey, implementation and application of preventive measures. *Vet Ital* 44: 469-481.

Monini M, Di Bartolo I, Ianiro G, Angeloni G, Magistrali CF, Ostanello F, Ruggeri FM. 2015. Detection and molecular characterization of zoonotic viruses in swine fecal samples in Italian pig herds. *Arch Virol* 160: 2547-2556.

Muscillo M, Fratini M, Graffeo R, Sanguinetti M, Martella V, Green KY, Della Libera S, La Rosa G. GIV 2013. Noroviruses in wastewaters and in stool specimens from hospitalized patients. *Food Environ Virol* 5:194–202.

Pavoni E, Consoli M, Suffredini E, Arcangeli G, Serracca L, Battistini R, Rossini I, Croci L, Losio MN. 2013. Noroviruses in seafood: a 9-year monitoring in Italy. *Foodborne Pathog Dis* 10: 533-539.

Pelosi E, Lambden PR, Caul EO, Liu B, Dingle K, Deng Y, Clarke IN. 1999. The seroepidemiology of genogroup 1 and genogroup 2 Norwalk-like viruses in Italy. *J Med Virol* 58: 93-99.

Pepe T, Ventrone I, Suffredini E, Ceruso M, Croci L, Anastasio A, Cortesi ML. 2012. Norovirus monitoring in bivalve molluscs harvested and commercialized in southern Italy. *J Food Prot* 75: 976-981.

Petrinca AR, Donia D, Pierangeli A, Gabrieli R, Degener AM, Bonanni E, Diaco L, Cecchini G, Anastasi P, Divizia M. 2009. Presence and environmental circulation of enteric viruses in three different wastewater treatment plants. *J Appl Microbiol* 106:1608-1617.

Prato R, Lopalco PL, Chironna M, Barbuti G, Germinario C, Quarto M. 2004. Norovirus gastroenteritis general outbreak associated with raw shellfish consumption in south Italy. *BMC Infect Dis* 4: 37.

Purpari G, Giammanco GM, Ruggeri FM, Rotolo V, Costantino C, Macaluso G, Vitale F, Spoto V, Geraci G, Mammina C, Di Bartolo I, Guercio A. 2012. Surveillance of a municipal drinking-water supply after a Norovirus outbreak in Italy. *Int J Infect Dis* 16S: e143.

Rimoldi SG, Stefani F, Pagani C, Chenal LL, Zanchetta N, Di Bartolo I, Lombardi A, Ruggeri FM, Di Lillo D, Zuccotti GV, Gismondo MR. 2011. Epidemiological and clinical characteristics of pediatric gastroenteritis associated with new viral agents. *Arch Virol*. 156: 1583-1589.

Rizzo C, Di Bartolo I, Santantonio M, Coscia MF, Monno R, De Vito D, Ruggeri FM, Rizzo G. 2007. Epidemiological and virological investigation of a Norovirus outbreak in a resort in Puglia, Italy. *BMC Infect Dis* 7: 135.

Rovida F, Campanini G, Sarasini A, Adzasehoun KMG, Piralla A, Baldanti F. 2007. Comparison of immunologic and molecular assays for the diagnosis of gastrointestinal viral infections. *Diag Microbiol Infect Dis* 75: 110-111

Scarcella C, Carasi S, Cadoria F, Macchi L, Pavan A, Salamana M, Alborali GL, Losio MM, Boni P, Lavazza A, Seyler T. 2009. An outbreak of viral gastroenteritis linked to municipal water supply, Lombardy, Italy, June 2009. *Euro Surveill* 14: pii=19274.

Serracca L, Rossini I, Battistini R, Goria M, Sant S, De Montis G, Ercolini C. 2012. Potential risk of norovirus infection due to the consumption of "ready to eat" food. *Food Environ Virol* 4: 89-92.

Serracca L, Verani M, Battistini R, Rossini I, Carducci A, Ercolini C. 2010. Evaluation of Adenovirus and *Escherichia coli* as indicators for human enteric viruses presence in mussels produced in La Spezia Gulf (Italy). *Lett Appl Microbiol* 50: 462-467.

Suffredini E, Corrain C, Arcangeli G, Fasolato L, Manfrin A, Rossetti E, Biazzi E, Mioni R, Pavoni E, Losio MN, Sanavio G, Croci L. 2008. Occurrence of enteric viruses in shellfish and relation to climatic-environmental factors. *Lett Appl Microbiol* 47: 467-74.

Suffredini E, La Rosa G, Proroga YTR, Della Libera S, Gregori G, Capuano F, De Medici D. 2016. Noroviruses in bivalve shellfish, in production waters and at discharge points: molecular characterization and source attribution. In: 5th Food and Environmental Virology Conference (ISFEV 2016). Proceedings; September 13-16, 2016; Kusatsu. 2016. p.174-176.

Suffredini E, Serracca L, Ercolini C, Cozzi L, Croci L. 2013. Ostriche responsabili di gastroenterite da Norovirus: valutazione dei livelli di contaminazione nel tempo in lotti provenienti dalla stessa area. In: II Convegno Nazionle Società Italiana di Ricerca Applicata alla Molluschicoltura. Proceedings; November 28-28, 2013; Cesenatico (Italy). p. 56-57.

Suffredini E, Lanni L, Arcangeli G, Pepe T, Mazzette R, Ciccaglioni G, Croci L. 2014. Qualitative and quantitative assessment of viral contamination in bivalve molluscs harvested in Italy. *Int J Food Microbiol* 184: 21-26.

Suffredini E, Magnabosco C, Civettini M, Rossetti E, Arcangeli G, Croci L. 2012. Norovirus contamination in different shellfish species harvested in the same production areas. *J Appl Microbiol* 113: 686-692.

Terio V, Martella V, Moschidou P, Di Pinto P, Tantillo G, Buonavoglia C. 2010. Norovirus in retail shellfish. *Food Microbiol* 27: 29-32.

Verani M, Casini B, Battistini R, Pizzi F, Rovini E., Carducci A. 2006. One-year monthly monitoring of Torque teno virus (TTV) in river water in Italy. *Water Sci Technol* 54: 191-195.

Wyn-Jones AP, Carducci A, Cook N, D'Agostino M, Divizia M, Fleischer J, Gantzer C, Gawler A, Girones R, Höller C, de Roda Husman AM, Kay D, Kozyra I, López-Pila J, Muscillo M, Nascimento MS, Papageorgiou G, Rutjes S, Sellwood J, Szewzyk R, Wyer M. 2011. Surveillance of adenoviruses and noroviruses in European recreational waters. *Water Res* 45: 1025-1038.

In: Noroviruses
Editor: Jesús L. Romalde
ISBN: 978-1-53611-939-8

Chapter 3

NOROVIRUSES IN DOGS

João Rodrigo Mesquita[1,2] and Maria São José Nascimento[3,*]

[1]Department of Zootechnics, Rural Engineering and Veterinary, Agrarian Superior School of the Polytechnic Institute of Viseu , Viseu, Portugal
[2]Centre for the Study of Education, Technologies and Health, Viseu, Portugal
[3]Department of Biological Sciences, Faculty of Pharmacy of the University of Porto, Porto, Portugal

ABSTRACT

Although known in humans since the early 1970s, it was not until 2008 that noroviruses were first described in dogs. Since the discovery in Italy, canine noroviruses were also found in dogs of Portugal, Greece, Hong Kong, USA and Japan, showing that this virus has worldwide circulation. This chapter describes the history, epidemiology and zoonotic potential of canine norovirus.

* Corresponding Author Address: Department of Biological Sciencies. Faculty of Pharmacy. Universty of Porto.4050-313 Porto, Portugal; E-mail: saojose@ff.up.pt.

Keywords: canine norovirus, epidemiology, zoonosis

INTRODUCTION

Noroviruses are a genetically diverse group of viruses belonging to the genus *Norovirus* of the family *Caliciviridae* that comprises four more genera, *Sapovirus, Lagovirus*, *Vesivirus* and *Nebovirus* (Green, 2007). All the viruses of this family are small icosahedric non-enveloped RNA viruses with 27 to 40 nm in diameter, with a positive-sense single stranded RNA of 7.5 to 7.7 kb (Green, 2000).

Norovirus genome contains 3 open reading frames (ORF1-3) with ORF1 encoding a polyprotein that is post-translationally cleaved into nonstructural proteins (NS1 to NS7) that are involved in viral replication, ORF2 encodes for the major structural protein (VP1) and ORF3 encodes a minor structural protein (VP2) (Green, 2007).

While murine norovirus can be cultured sucessfully this is not the case for the rest of noroviruses, including human noroviruses (Ettayebi et al., 2016). Noroviruses are today classified in five genogroups (GI to GV), with subdivision into clusters and strains, based on the complete amino acid sequence of VP1 (Zheng et al., 2005) but two additional genogroups (GVI and GVII) have recently been proposed (Vinjé, 2014). GI, GII, GIV and GVI are known to infect humans, being GII the most prevalent (Zheng et al., 2005; Glass et al., 2009). Noroviruses belonging to GIV and GVI were described in dogs (Martella et al., 2008; Mesquita et al., 2010), however recently discovered strains were proposed to belong to GVII (Vinjé, 2014).

THE HISTORY OF CANINE NOROVIRUS

Although known in humans since the early 1970s (Kapikian et al., 1972), it was not until 2008 that noroviruses were described in dogs for the first time (Martella et al., 2008). But caliciviruses had been detected in dogs long before. In 1979 the first canine calicivirus was found having been named canine glossitis virus (strain WSU 79-1831) (Evermann et al., 1981). Researchers found that the canine glossitis virus grew in Crandell feline kidney cells, and electron microscopic studies allowed ascertaining that the virion structure was consistent with caliciviruses (Evermann et al., 1983). At that time no genomic characterization techniques were available and viral characterization was achieved mainly by electron microscopy studies. Since then several calicivirus-like particles have been occasionally identified by electron microscopy in fecal specimens from dogs with diarrhea and, in some instances, in dogs with glossitis, balanitis, and vesicular vaginitis, however most isolates were found to be feline caliciviruses and were likely acquired from cats (Evermann et al., 1985; Crandell, 1988; San Gabriel et al., 1997; Martella et al., 2002).

Thus far, only two reports have described the identification of authentic caliciviruses (non-norovirus) in dogs (Schaffer et al., 1985; Mochizuki et al., 1993). In 1985 a calicivirus was isolated in cell culture from the stools of a 4-year-old dog with bloody diarrhea and central nervous system disturbance in Tennessee, USA (Schaffer et al., 1985). This virus was found to replicate in experimentally infected dogs inducing seroconversion showing also to be easily propagated in cell culture of canine and dolphin cell lines (Schaffer et al., 1985). It was antigenically unrelated to feline caliciviruses but it was classified as a calicivirus based on its morphology and physicochemical properties (density, sedimentation rate, single major polypeptide, RNA genome size). It showed to be distantly related to previously described caliciviruses and closer to the vesiviruses, not reacting with antisera to numerous caliciviruses except with the stunting syndrome agent of chickens (Schaffer et al., 1985). Antibodies against this virus were identified in 76% of the canine sera studied (Schaffer et al., 1985). In 1990, another calicivirus was identified in Japan in the stools of a 2-month-old pup with intermittent watery diarrhea (Mochizuki et al., 1993). The virus (named strain 48) was also found to be antigenically and genetically unrelated to feline calicivirus and was tentatively proposed as a canine calicivirus (Roerink et al., 1999;

Matsuura et al, 2002). Antibodies against strain 48 were detected in 57% of dogs in Japan (Mochizuki et al., 2002) and in 36.5% of dogs in Korea (Jang et al., 2003).

It will be only in 2008 that the first norovirus is discovered in dogs, namely in the stools of a 60-day-old mixed-breed pup with diarrhea and vomit in Italy (Martella et al., 2008). This new norovirus was genetically characterized and found to be similar to GIV noroviruses found in humans (Martella et al., 2008). Researchers found difficult to speculate on the pathogenic role of this norovirus because a canine parvovirus-2a was also present in the pup stools, which was probably the agent responsible for the severity of the observed clinical signs. Nevertheless the norovirus was detected in the stools of the pup for 22 days, suggesting an active viral replication, rulling out the possibility of passive viral transit (mechanical passage) in the dog intestinal tract (Martella et al., 2008). Soon after, in 2010, norovirus sequences were detected in stools of 40% (25 out of 63 stools) of dogs with diarrhea but only in 9% (4 out of 42 stools) of dogs without diarrhea in Portugal (Mesquita et al., 2010). Although canine coronavirus and parvovirus 2 were also detected in some dog stools it was possible to find a statistical association between diarrhea and canine norovirus as well as with canine coronavirus, but not canine parvovirus 2 (Mesquita et al., 2010). All the norovirus sequences detected in the Portuguese dogs presented a high nucleotide sequence identity (range 98% – 100%) but differed 19% – 22% from the canine norovirus strains detected previously in Italy (Mesquita et al., 2010). This new norovirus was genetically characterized and tentatively classified as a GVI norovirus and named Viseu strain (Mesquita et al., 2010). At that time another detection of a canine norovirus was reported in Greece during the course of a pilot study to investigate the presence of norovirus in dogs with diarrheic symptoms (N = 72) from kennels and pet shops as well as in animals hospitalized in veterinary clinics (Ntafis et al., 2010). The canine norovirus detected in Greece was found to be geneticaly similar to the canine norovirus GIV first detected in Italy (Ntafis et al., 2010).

Since then, canine noroviruses were detected also in the United States (Azevedo et al., 2012) and in Hong Kong (Tse et al., 2012) where, as part of a surveillance study of animal viruses, a novel norovirus strain was described in two canine fecal swab specimens and found to be distantly related to other noroviruses, including other canine noroviruses, potentially belonging to a novel genogroup (GVII) (Vinjé, 2014). Canine noroviruses were also detected in two dogs in Japan in a study on stools of household dogs (N = 97), with partial sequence analysis showing highest homology to GIV canine norovirus strains (Soma et al., 2015).

WHAT IS KNOWN OF ITS EPIDEMIOLOGY?

The epidemiological features of canine norovirus infection are still poorly documented and rely on a scarce number of studies documenting outbreaks, serosurveys and prospective molecular epidemiological studies. In an outbreak of acute gastroenteritis in dogs that occurred in March 2011 in a kennel in Lisbon, Portugal, some epidemiological features were registered (Mesquita and Nascimento, 2012a). Five French Bulldogs that were housed in this kennel, developed diarrhea 48h after the arrival of two diarrheic dogs coming from Moscow, Russia. These two Russian dogs travelled from Moscow to Lisbon through distinct routes and transports, namely by road and by plane. They had been bred separately by distinct Russian dog breeders and have never been in contact before their introduction to the Portuguese kennel. As part of the purchase agreement, both Russian dogs were fully dewormed and vaccinated against canine distemper virus, canine adenovirus-2, canine parvovirus, canine parainfluenza virus and rabies before the international shipping, but at their arrival to the Lisbon kennel the two dogs presented diarrhea. Canine noroviruses were detected in the stools of all the diarrheic dogs and the nucleotide sequence analysis showed that all canine norovirus strains were identical, presenting highest homology to the canine norovirus GVI detected in Portugal (Mesquita and Nascimento, 2012a). From this canine norovirus outbreak it was possible to identify some epidemiological features that curiously resembled those of human norovirus infections. The canine norovirus infection disseminated rapidly in 2 days to all dogs in the kennel demonstrating the highly contagious nature of this virus, the incubation period was <48 h, the diarrhoeal disease was self-limiting and the viral shedding lasted less than 7 days (Mesquita and Nascimento, 2012a). This had not been the only canine norovirus outbreak reported as another epidemic of norovirus infection was identified in a kennel in Greece (Ntafis et al., 2010). The analysis of nucleotide sequences showed that all dogs of this epidemic were infected with the same strain and the clonal origin of the strains was consistent with a contamination from a common source. The Greek epidemic also showed that canine norovirus can easily spread within a community of dogs, i.e., in its homologous host, demonstrating also the high infectivity of this norovirus (Ntafis et al., 2010).

In a later study performed by our team in Portugal the prevalence of canine norovirus fecal excretion in a 4-year period was accessed (Mesquita and Nascimento, 2012b). From 2007-2011, 256 stools were collected from dogs across Portugal and tested for the presence of canine norovirus. An overall prevalence of norovirus fecal shedding of 23% (60/256) was found. All sequences had a high nucleotide identity (range 98%-100%) to the canine norovirus first described in Portugal (Viseu strain). Interestingly, the shedding rate peak was detected during the winter months, a well-known epidemiologic feature of human norovirus infections.

Further epidemiological studies of norovirus infection in dogs have also been performed in the United States (Azevedo et al., 2012), UK (Caddy et al., 2013) and Portugal (Mesquita et al., 2014). The study in the United States described a prevalence of 11% in dog diarrhoeic samples. The UK study investigated the prevalence of canine norovirus not only by screening canine stool samples for the presence of canine norovirus (N = 228), but also by testing dog serum samples (N = 396) for the presence of anti-canine norovirus antibodies by ELISA. Although no stools were found positive for canine norovirus the seroprevalence screen revealed a high level of anti-canine norovirus seropositivity in dogs, namely in 38.1% of samples collected between 1999–2001 and in 60.1% of samples collected in 2012–2013 (Caddy et al., 2013). This increase in seroprevalence over time led authors to suggest that canine norovirus infections were becoming more widespread.

A Portuguese study broadened the geographic focus and investigated the prevalence of antibodies against canine noroviruses (Viseu strain) in serum samples collected from dogs that visited veterinary clinics in 14 countries in Europe. From the 510 dog serum samples collected, 183 (36%) tested positive for antibodies against canine norovirus, namely in dogs from Denmark (10 positive out of 50), Finland (7 positive out of 10), France (10 positive out of 50), Germany (8 positive out of 50), Italy (6 positive out of 10), Netherlands (17 positive out of 50), Norway (16 positive out of 50), Poland (16 positive out of 50), Portugal (64 positive out of 100), Sweden (4 positive out of 10), Switzerland (2 positive out of 10) and the United Kingdom (23 positive out of 50). This study showed the widespread circulation of norovirus in dogs in Europe (Mesquita et al., 2014).

Are Noroviruses Potential Zoonotic Agents?

Throughout the years acumulating evidence is building and pointing to the possibility of canine noroviruses transmission to humans, and human noroviruses transmission to dogs. Initial studies have shown high prevalence of antibodies against human norovirus (Norwalk strain) both in pigs and captive juvenile macaques suggesting that human norovirus could infect animals (Farkas et al., 2005; Farkas et al., 2010). On the other hand, in the Netherlands, an increased exposure risk to bovine norovirus was demonstrated for large animal veterinarians since they presented a higher seroprevalence of bovine norovirus antibodies compared to the general population, suggesting that animal noroviruses infect humans (Widdowson et al., 2005). More recently, human noroviruses GII were found in faecal samples from pet dogs that had been in direct contact with symptomatic persons in Finland (Summa et al., 2012). Although replication of human noroviruses in dogs was not definitely shown, results suggested that at least these viruses survived in the canine gastrointestinal tract (Summa et al., 2012).

In a study conducted by our team (Mesquita et al., 2013) that aimed to determine the zoonotic potential of canine norovirus, sera from pet veterinarians (N = 373) and age-matched general population controls (N = 120) were screened for the presence of antibodies to virus-like particles of canine norovirus (Viseu strain). Antibodies against canine norovirus were found in 22.3% of pet veterinarians and in 5.8% of the general population control suggesting that pet veterinarians are at an increased risk for exposure and possibly infection with this virus. The presence of antibodies against canine norovirus in the population control was explained by household contact with dogs, as they are popular pet animals in the Portuguese population (Mesquita et al., 2013).

A more recent study on the zoonotic risk of transmission of human norovirus to dogs, an UK team studied the ability of canine tissues to bind to human norovirus in vitro (Caddy et al., 2015). Results showed that seven different genotypes of human norovirus (GI.1, GI.2, GI.3, GII.3, GII.4, GII.6, and GII.12) virus-like particles (VLPs) were able to bind to canine gastrointestinal tissue, demonstrating the possibility of infection (Caddy et al., 2015). They also analyzed dog stools for the presence of human norovirus and dog sera for the presence of anti-human norovirus antibodies. Although no human norovirus was detected in the dog stools, antibodies against human norovirus were observed in 43 of the 325 dog serum samples tested, which demonstrates previous exposure to human norovirus. Curiously, canine seroprevalence for the seven different human norovirus genotypes mirrored the seroprevalence in the human population. Although in this study entry and replication within cells had not been demonstrated, the canine serological data indicated that dogs produce an immune response to human norovirus, implying productive infection and revealing zoonotic implications for human norovirus (Caddy et al., 2015).

Altogether, data from these studies suggest that noroviruses may be able to cross the species barrier and raise alerts for potential cross-species transmission events.

References

Azevedo M., Mullis L., Vegas E., Britt J., Pereira O. 2012. Detection of norovirus in dogs in Arkansas. *American Society of Virology Conference*. 23–30.

Caddy S., Emmott E., El-Attar L., Mitchell J., de Rougemont A., Brownlie J., Goodfellow I. 2013. Serological evidence for multiple strains of canine norovirus in the UK dog population. *PLoS One* 8: e81596.

Caddy S. L., de Rougemont A., Emmott E., El-Attar L., Mitchell J. A., Hollinshead M., Belliot G., Brownlie J., Le Pendu J., Goodfellow I. 2015. Evidence for human norovirus infection of dogs in the United Kingdom. *J. Clin. Microbiol.* 53: 1873-83.

Crandell R. A. 1988. Isolation and characterization of caliciviruses from dogs with vesicular genital disease. *Arch. Virol.* 98: 65-71.

Ettayebi K., Crawford S. E., Murakami K., Broughman J. R., Karandikar U., Tenge V. R., Neill F. H., Blutt S. E., Zeng X. L., Qu L., Kou B., Opekun A. R., Burrin D., Graham D. Y., Ramani S., Atmar R. L., Estes M. K. 2016. Replication of human noroviruses in stem cell-derived human enteroids. *Science* 353: 1387-1393.

Evermann J. F., Bryan G. M., McKiernan A. J. 1981. Isolation of a calicivirus from a case of canine glossitis. *Canine Pract.* 8: 36-39.

Evermann J. F., McKeirnan A. J., Smith A. W. 1985. Isolation and identification of Caliciviruses from dogs with enteric infections. *Am. J. Vet. Res.* 46: 218-20.

Evermann J. F., Smith A. W., Skilling D. E., McKeirnan A. J. 1983. Ultrastructure of newly recognized Caliciviruses of the dog and mink. *Arch. Virol.* 76: 257-61.

Farkas T., Dufour J., Jiang X., Sestak K. 2010. Detection of norovirus-, sapovirus- and rhesus enteric calicivirus-specific antibodies in captive juvenile macaques. *J. Gen. Virol.* 91: 734-738.

Farkas T., Nakajima S., Sugieda M., Deng X., Zhong W., Jiang X. 2005. Seroprevalence of noroviruses in swine. *J. Clin. Microbiol.* 43: 657-661.

Glass R. I., Parashar U. D., Estes M. K. 2009. Norovirus gastroenteritis. *N. Engl. J. Med. 361*: 1776–1785.

Green K. Y., Ando T., Balayan M. S., Berke T., Clarke I. N., Estes M. K., Matson D. O., Nakata S., Neill J. D., Studdert M. J., Thiel H. J. 2000. Taxonomy of the caliciviruses. *J. Infect. Dis.* 181: S322-S330.

Green K. Y. 2007. Caliciviridae: The Noroviruses. In: Knipe DM, Howley PM. (Eds.) Fields Virology, 6th ed., pp. 949-979. *Lippincott Williams and Wilkins*. Philadelphia, PA.

Jang H. K., Tohya Y., Han K. Y. 2003. Seroprevalence of canine calicivirus and canine minute virus in the Republic of Korea. *Vet. Rec.* 153: 150-2.

Martella V., Lorusso E., Decaro N., Elia G., Radogna A., D'Abramo M., Desario C., Cavalli A., Corrente M., Camero M., Germinario C. A., Bányai K., Di Martino B., Marsilio F., Carmichael L. E., Buonavoglia C. 2008. Detection and molecular characterization of a canine norovirus. *Emerg. Infect. Dis.* 14: 1306–8.

Martella V., Pratelli A., Gentile M. 2002. Analysis of the capsid protein gene of a feline-like calicivirus isolated from a dog. Vet. *Microbiol.* 8: 315-22.

Matsuura Y., Tohya Y., Nakamura K. 2002. Complete nucleotide sequence, genome organization and phylogenic analysis of the canine calicivirus. *Virus Genes* 25: 67-73.

Mesquita J. R., Barclay L., Nascimento M. S. J., Vinjé J. 2010. Novel norovirus in dogs with diarrhea. *Emerg. Infect. Dis.* 16: 980-982.

Mesquita J. R., Costantini V. P., Cannon J. L., Lin S. C., Nascimento M. S., Vinjé J. 2013. Presence of antibodies against genogroup VI norovirus in humans. *Virol. J.* 10: 176.

Mesquita J. R., Delgado I., Costantini V., Heenemann K., Vahlenkamp T. W., Vinjé J., Nascimento M. S. 2014. Seroprevalence of canine norovirus in 14 European countries. *Clin. Vaccine Immunol.* 21: 898-900.

Mesquita J. R., Nascimento M. S. J. 2012a. Gastroenteritis outbreak associated with faecal shedding of canine norovirus in a Portuguese kennel following introduction of imported dogs from Russia. *Transbound Emerg. Dis.* 2012a 59: 456-459.

Mesquita J. R., Nascimento M. S. 2012b. Molecular epidemiology of canine norovirus in dogs from Portugal, 2007-2011. *BMC Vet. Res.* 8: 107.

Mochizuki M., Hashimoto M., Roerink F. 2002. Molecular and seroepidemiological evidence of canine calicivirus infections in Japan. *J. Clin. Microbiol.* 40: 2629-2631.

Mochizuki M., Kawanishi A., Sakamoto H. 1993. A calicivirus isolated from a dog with fatal diarrhoea. *Vet. Rec.* 132: 221-222.

Ntafis V., Xylouri E., Radogna A., Buonavoglia C., Martella V. 2010. Outbreak of canine norovirus infection in young dogs. *J. Clin. Microbiol.* 48: 2605-2608.

Roerink F., Hashimoto M., Tohya Y. 1999. Organization of the canine calicivirus genome from the RNA polymerase gene to the poly(A) tail. *J. Gen. Virol.* 80: 929-935.

San Gabriel MC, Tohya Y, Sugimura T. 1997. Identification of canine calicivirus capsid protein and its immunoreactivity in western blotting. *J. Vet. Med. Sci.* 59: 97-101.

Schaffer F. L., Soergel M. E., Black J. W. 1985. Characterization of a new calicivirus isolated from feces of a dog. *Arch. Virol.* 84: 181-195.

Soma T., Nakagomi O., Nakagomi T., Mochizuki M. 2015. Detection of Norovirus and Sapovirus from diarrheic dogs and cats in Japan. *Microbiol. Immunol.* 59: 123–128.

Summa M. von Bonsdorff C. H., Maunula L. 2012. Pet dogs-a transmission route for human noroviruses? *J. Clin. Virol.* 53: 244–247.

Teunis, P. F., Moe C. L., Liu P., Miller S. E., Lindesmith L., Baric R. S., Le Pendu J., Calderon R. L. 2008. Norwalk virus: how infectious is it? *J. Med. Virol.* 80: 1468–1476.

Tse H., Lau S. K. P., Chan W-M, Choi G. K. Y., Woo P. C. Y., Yuen K. Y. 2012. Complete genome sequences of novel canine noroviruses in Hong Kong. *J. Virol.* 86: 9531–9532.

Vinjé J. 2014. Advances in Laboratory Methods for Detection and Typing of Norovirus. *J. Clin. Microbiol.* 53: 373-381.

Widdowson M. A., Rockx B., Schepp R., van der Poel W. H., Vinje J., van Duynhoven Y. T., Koopmans M. P. 2005. Detection of serum antibodies to bovine norovirus in veterinarians and the general population in the Netherlands. *J. Med. Virol.* 76: 119-128.

Zheng D. P., Ando T., Fankhauser R. L., Beard R. S., Glass R. I., Monroe S. S. 2005. Norovirus classification and proposed strain nomenclature. *Virol.* 15: 312–323.

In: Noroviruses
Editor: Jesús L. Romalde

ISBN: 978-1-53611-939-8

Chapter 4

BAYESIAN COALESCENT INFERENCE DEMONSTRATES DIFFERENCES IN EVOLUTIONARY DYNAMICS AMONG NOROVIRUS GII GENOTYPES

***Carmen F. Manso*[*] *and Jesús L. Romalde*[†]**
Departamento de Microbiología y Parasitología,
CIBUS-Facultad de Biología, Universidade de Santiago de Compostela,
Santiago de Compostela, Spain

ABSTRACT

The high incidence of Norovirus (NoV) infections seems to be related with a pattern of continuous lineage turnover of NoV GII.4, with emerging new variants replacing those previously predominant. It has been proposed that other NoV genotypes evolved similarly to GII.4,

---

[*] Corresponding Author address: Antiviral Unit. Virus Reference Department. Public Health England. 61 Colindale Ave. London NW9 5EQ. UK.
[†] Corresponding Author address: Departamento de Microbiología y Parasitología, CIBUS-Facultad de Biología. Universidade de Santiago de Compostela. Campus Vida s/n. 15782, Santiago de Compostela. Spain. Email: jesus.romalde@usc.es.

suggesting the future emergence and widespread of other epidemic genotype variants. In this work, evolutionary rates and population dynamics of different genotypes of NoV GII were inferred by Bayesian coalescent approach. This study provides an insight into the complex evolutionary patterns of different NoV GII genotypes and reports for the first time the evolutionary rates of GII.7, GII.13 and GII.14. The estimated mean evolutionary rate for NoV GII.14 was 10^{-2} substitution/site/year. For the other NoV genotypes included in the study, GII.3, GII.7 and GII.13, the estimated mean evolutionary rates were lower, with values of 10^{-3} substitution/site/year. The MCC (maximum clade credibility) tree, showing a time measured phylogenetic relationship among the different genotypes analysed, revealed that GII.4 and GII.14 are the most ancient genotypes, diverging respectively from common TMRCAs (Time to Most Recent Common Ancestor) dated 83 and 54 years ago. Results obtained for GII.4 BSP (Bayesian Skyline Plot) revealed an increment in $N_e\tau$ values coincident with the appearance of 2006 and 2009 variants. GII.14 BSP shows a high increase in $N_e\tau$ values from year 2009, with peaks in 2010 and 2011. The high evolutionary rate of NoV GII.14, together with its observed population dynamics might be correlated with the current widespread of new genotype variants.

Keywords: norovirus, evolution, genotype, Bayesian analysis

INTRODUCTION

Noroviruses (NoVs) are the foremost cause of non-bacterial acute gastroenteritis (Glass et al., 2009; Siebenga et al., 2009). NoV transmissions occur by fecal oral route, either by direct contact person to person or via contaminated surfaces, food or water. Illness is usually self-limiting and symptoms, comprising acute onset vomiting and watery diarrhea, commonly persist from one to three days (Rockx et al., 2002). The relevance of NoV studies lays in their high prevalence in the population (de Wit et al., 2001) and in the more severe and prolonged illness in elderlies and immunosuppressed patients (Okada et al., 2006; Harris et al., 2008; Siebenga et al., 2008). NoVs are extremely contagious, owing to a very low infectious dose and to high levels of shedding (Chan et al., 2006; Teunis et al., 2008; Tu et al., 2008). Additionally, although patients usually recover from symptoms in 2-3 days, high levels of shedding may last several weeks (Tu et al., 2008). No or limited long-term immunity results from infection, so one person may be repeatedly infected (Johnson et al., 1990).

Noroviruses, classified in the family *Caliciviridae*, are a group of nonenveloped, icosahedral viruses, with a single-stranded positive sense RNA genome. NoVs are genetically diverse, being classify into 33 genetic clusters within 6 genogroups (GI-GVI), of which GI and GII strains affect most commonly humans (Green et al., 1994; Wang et al., 1994; Ando et al., 2000; Patel et al., 2009). Genogroups I, II and rarely IV are responsible for human infections (Zheng et al., 2011). Within the genogroup II, strains belonging to genotype II.4 are the most common cause of gastroenteritis outbreaks, including pandemics, and sporadic cases since the mid-1990s (Siebenga et al., 2009; Leon et al., 2011).

The high incidence of NoV infections seems to be related to the emergence of new GII.4 variants that had evolved by genetic drift of the capsid gen or homologous recombination (Siebenga et al., 2007; Motomura et al., 2008; Bull et al., 2011; Eden et al., 2013; Medici et al., 2014; Sabriá et al., 2014). The major capsid protein, VP1, contains sites that are predicted to interact with Histo-Blood Group Antigens (HBGAs), and mutations occurring in this region may be driven by evasion of the host immune response (Donaldson et al., 2008). The pattern of continuous lineage turnover, referred to as epochal evolution, with emerging new variants replacing previously predominant circulating ones is strongly reminiscent of what is observed in the molecular epidemiology of influenza A virus (IAV) (Lindesmith et al., 2008).

Occasional increases in prevalence of other genotypes aside from GII.4 have been reported (Lewis et al., 1997; Koopmans et al., 2000; Hoffmann et al., 2010; Iritani et al., 2010; Sabriá et al., 2014) and it has been proposed that these NoV genotypes may evolved similarly to GII.4, suggesting the future emergence and widespread of new epidemic variants (Iritani et al., 2008). However, little is known about the evolutionary rates, viral spread and population dynamics of NoV populations. Other NoV genotypes like GII.3 and GII.7 have been proposed to present an evolutionary disadvantage compared to GII.4 due to a lower rate of evolution and a less processive polymerase (Bull et al., 2010).

In order to get insight into the evolutionary mechanism of different NoV GII genotypes, a total 306 partial capsid gene sequences obtained from strains detected in Spain were studied together with other sequences from strains detected in different countries with known detection dates. Bayesian coalescent approach was employed to infer the evolutionary rates and population dynamics of different NoV GII genotypes. The time to the most recent common ancestor (TMRCA) of Spanish strains belonging to different NoV GII genotypes could also be inferred by this approach.

METHODS

Sequence Data

A total of 306 partial capsid gene sequences were newly acquired as part of a NoVs prevalence study during a one-year period (July 2010-June 2011) in NW Spain (Manso and Romalde, 2013). Sequences were deposited in the GenBank database and their accession numbers are shown in Table 1. Genotyping of the detected NoV strains was carried out by semi-nested RT-PCR protocol with specific primers for NoV GII. Phylogenetic analyses revealed that 139 sequences belonged to GII.4 and 107 belonged to GII.14. Twenty-one of the sequences were characterized as GII.7, 21 as GII.13, 6 as GII.3, 6 as GII.12 and 2 as GII.1 (Manso and Romalde, 2013).

Partial capsid gene sequences with known detection year were retrieved from public databases Genbank and Noronet (www.noronet.nl) corresponding to genotypes GII.4 (74 sequences), GII.14 (16 sequences), GII.7 (26 sequences), GII.13 (31 sequences) and GII.3 (77 sequences). Not enough sequences corresponding to the target region were found in the databases to perform the coalescent analysis in the case of GII.1 and GII.12. Accession numbers of GenBank sequences and outbreak number of Noronet sequences are shown in Table 2.

Sequences were aligned with MEGA5 software (Tamura et al., 2011) and trimmed to a common length of 273 nt. Final datasets consisted in 213 sequences of NoV GII.4 comprising a 37-year period, 123 sequences of GII.14 comprising a 12-year period, 47 sequences of GII.7 over a 21-year period, 82 sequences over a 36-year period for GII.3 and 52 sequences of GII.13 over a 13-year period.

Table 1. Accession numbers of the partial capsid sequences obtained in the present work

Strain	Accession Number	Strain	Accession Number	Strain	Accession Number
NoV GII.4					
H88G2010.08	LN482637	H1609G2011.02	LN482728	H1321G2011.01	LN482706
H98G2010.08	LN482638	H1730G2011.02	LN482729	H1331G2011.01	LN482707
H115G2010.08	LN482639	H1842G2011.03	LN482730	H1346G2011.01	LN482708
H165G2010.08	LN482640	H1847G2011.03	LN482731	H1363G2011.01	LN482709
H171G2010.08	LN482641	H1848G2011.03	LN482732	H1369G2011.01	LN482710
H174G2010.08	LN482642	H1849G2011.03	LN482733	H1384G2011.01	LN482711
H176G2010.08	LN482643	H1930G2011.03	LN482734	H1405G2011.01	LN482712
H233G2010.08	LN482644	H1960G2011.03	LN482735	H1407G2011.01	LN482713
H255G2010.08	LN482645	H1961G2011.03	LN482736	H1459G2011.02	LN482714
H326G2010.09	LN482646	H1968G2011.03	LN482737	H1462G2011.02	LN482715
H375G2010.09	LN482647	H1974G2011.03	LN482738	H1478G2011.01	LN482716
H382G2010.09	LN482648	H1975G2011.03	LN482739	H1483G2011.02	LN482717
H383G2010.09	LN482649	H1976G2011.03	LN482740	H1493G2011.02	LN482718
H385G2010.09	LN482650	H1980G2011.03	LN482741	H1499G2011.02	LN482719
H419G2010.09	LN482651	H1981G2011.03	LN482742	H1511G2011.02	LN482720
H489G2010.09	LN482652	H2020G2011.04	LN482743	H1523G2011.02	LN482721
H498G2010.10	LN482653	H2023G2011.04	LN482744	H1529G2011.02	LN482722
H534G2010.10	LN482654	H2028G2011.04	LN482745	H1531G2011.02	LN482723
H535G2010.10	LN482655	H2057G2011.04	LN482746	H1553G2011.01	LN482724
H545G2010.10	LN482656	H2080G2011.04	LN482747	H2114G2010.04	LN482750
H554G2010.10	LN482657	H2104G2011.04	LN482748	H2151G2011.04	LN482751
H562G2010.10	LN482658	H2109G2011.04	LN482749	H2207G2011.04	LN482752
H567G2010.10	LN482659	H929G2010.11	LN482681	H2219G2011.04	LN482753
H572G2010.10	LN482660	H970G2010.12	LN482682	H2220G2011.04	LN482754
H582G2010.10	LN482661	H982G2010.12	LN482683	H2222G2011.04	LN482755
H610G2010.10	LN482662	H1028G2010.12	LN482684	H2223G2011.04	LN482756
H615G2010.10	LN482663	H1045G2010.12	LN482685	H2225G2011.04	LN482757

Table 1. (Continued)

Strain	Accession Number	Strain	Accession Number	Strain	Accession Number
H622G2010.10	LN482664	H1052G2010.12	LN482686	H2226G2011.04	LN482758
H627G2010.10	LN482665	H1070G2010.12	LN482687	H2292G2011.04	LN482759
H641G2010.10	LN482666	H1174G2010.12	LN482688	H2295G2011.04	LN482760
H643G2010.10	LN482667	H1121G2010.12	LN482689	H2301G2011.04	LN482761
H648G2010.10	LN482668	H1141G2010.12	LN482690	H2317G2011.05	LN482762
H652G2010.10	LN482669	H1200G2010.12	LN482691	H2344G2011.05	LN482763
H666G2010.10	LN482670	H1203G2010.12	LN482692	H2345G2011.05	LN482764
H681G2010.10	LN482671	H1213G2011.01	LN482693	H2347G2011.05	LN482765
H685G2010.10	LN482672	H1220G2011.01	LN482694	H2348G2011.05	LN482766
H694G2010.10	LN482673	H1221G2011.01	LN482695	H2349G2011.05	LN482767
H755G2010.10	LN482674	H1226G2011.01	LN482696	H2351G2011.05	LN482768
H766G2010.11	LN482675	H1238G2011.01	LN482697	H2359G2011.05	LN482769
H790G2010.11	LN482676	H1243G2011.01	LN482698	H2365G2011.05	LN482770
H839G2010.11	LN482677	H1247G2011.01	LN482699	H2393G2011.05	LN482771
H898G2010.11	LN482678	H1249G2011.01	LN482700	H2551G2011.06	LN482772
H902G2010.11	LN482679	H1263G2011.01	LN482701	H2581G2011.06	LN482773
H917G2010.11	LN482680	H1267G2011.01	LN482702	H2590G2011.06	LN482774
H1575G2011.02	LN482725	H1273G2011.01	LN482703	H2596G2011.06	LN482775
H1578G2011.02	LN482726	H1284G2011.01	LN482704	H1321G2011.01	LN482706
H1587G2011.02	LN482727	H1294G2011.01	LN482705		
NoV GII.14					
H665G2010.10	LN482776	H1274G2011.01	LN482836	H1149G2010.12	LN482819
H670G2010.10	LN482777	H1279G2011.01	LN482837	H1152G2010.12	LN482820
H671G2010.10	LN482778	H1308G2011.01	LN482838	H1158G2010.12	LN482821
H706G2010.10	LN482779	H1314G2011.01	LN482839	H1164G2010.12	LN482822
H720G2010.10	LN482780	H1408G2011.01	LN482840	H1166G2010.12	LN482823
H744G2010.10	LN482781	H1458G2011.02	LN482841	H2083G2011.04	LN482853
H750G2010.10	LN482782	H1580G2011.02	LN482842	H2085G2011.04	LN482854
H782G2010.11	LN482783	H1771G2011.03	LN482843	H2086G2011.04	LN482855
H785G2010.11	LN482784	H1861G2011.03	LN482844	H2087G2011.04	LN482856
H794G2010.11	LN482785	H1944G2011.03	LN482845	H2088G2011.04	LN482857
H795G2010.11	LN482786	H2016G2011.04	LN482846	H2090G2011.04	LN482858
H798G2010.11	LN482787	H2031G2011.04	LN482847	H2100G2011.04	LN482859
H800G2010.11	LN482788	H2067G2011.04	LN482848	H2129G2011.04	LN482860
H801G2010.11	LN482789	H2073G2011.04	LN482849	H2132G2011.04	LN482861
H802G2010.11	LN482790	H2078G2011.04	LN482850	H2139G2011.04	LN482862
H803G2010.11	LN482791	H2079G2011.04	LN482851	H2159G2011.04	LN482863
H804G2010.11	LN482792	H2081G2011.04	LN482852	H2211G2011.04	LN482864
H805G2010.11	LN482793	H897G2010.11	LN482800	H2215G2011.04	LN482865

Strain	Accession Number	Strain	Accession Number	Strain	Accession Number
H806G2010.11	LN482794	H899G2010.11	LN482801	H2216G2011.04	LN482866
H864G2010.11	LN482795	H911G2010.11	LN482802	H2218G2011.04	LN482867
H870G2010.11	LN482796	H912G2010.11	LN482803	H2227G2011.04	LN482868
H889G2010.11	LN482797	H916G2010.11	LN482804	H2228G2011.04	LN482869
H893G2010.11	LN482798	H920G2010.11	LN482805	H2231G2011.04	LN482870
H896G2010.11	LN482799	H937G2010.11	LN482806	H2244G2011.04	LN482871
H1169G2010.12	LN482824	H939G2010.11	LN482807	H2263G2011.04	LN482872
H1178G2010.12	LN482825	H945G2010.11	LN482808	H2264G2011.04	LN482873
H1231G2011.01	LN482826	H987G2010.12	LN482809	H2266G2011.04	LN482874
H1236G2011.01	LN482827	H1031G2010.12	LN482810	H2270G2011.04	LN482875
H1239G2011.01	LN482828	H1036G2010.12	LN482811	H2437G2011.05	LN482876
H1258G2011.01	LN482829	H1039G2010.12	LN482812	H2476G2011.05	LN482877
H1260G2011.01	LN482830	H1046G2010.12	LN482813	H2510G2011.05	LN482878
H1261G2011.01	LN482831	H1048G2010.12	LN482814	H2609G2011.06	LN482879
H1264G2011.01	LN482832	H1069G2010.12	LN482815	H2631G2011.06	LN482880
H1268G2011.01	LN482833	H1140G2010.12	LN482816	H2639G2011.06	LN482881
H1269G2011.01	LN482834	H1147G2010.12	LN482817	H2657G2011.06	LN482882
H1271G2011.01	LN482835	H1148G2010.12	LN482818		
NoV GII.3					
H1026G2011.12	LN482883	H1043G2011.12	LN482885	H1409G2012.01	LN482887
H1042G2011.12	LN482884	H1106G2011.12	LN482886		
NoV GII.7					
H10G2010.07	LN482888	H758G2010.10	LN482895	H1819G2011.03	LN482902
H28G2010.07	LN482889	H796G2010.11	LN482896	H2107G2011.04	LN482903
H73G2010.08	LN482890	H797G2010.11	LN482897	H2110G2011.04	LN482904
H227G2010.08	LN482891	H815G2010.11	LN482898	H2180G2011.04	LN482905
H309G2010.09	LN482892	H829G2010.11	LN482899	H2490G2011.05	LN482906
H695G2010.10	LN482893	H1191G2010.12	LN482900	H2491G2011.05	LN482907
H701G2010.10	LN482894	H1681G2011.02	LN482901	H2492G2011.05	LN482908
NoV GII.13					
H18G2010.07	LN482909	H628G2010.10	LN482919	H799G2010.10	LN482924
H422G2010.09	LN482910	H629G2010.10	LN482920	H1022G2010.12	LN482925
H504G2010.10	LN482913	H632G2010.10	LN482921	H1023G2010.12	LN482926
H556G2010.10	LN482914	H480G2010.09	LN482911	H1194G2010.12	LN482927
H561G2010.10	LN482915	H481G2010.09	LN482912	H1196G2010.12	LN482928
H592G2010.10	LN482916	H664G2010.10	LN482922	H1563G2011.02	LN482929
H599G2010.10	LN482917	H689G2010.10	LN482923	H2617G2011.06	LN482930
H611G2010.10	LN482918				

Table 2. Accession numbers and source of the reference sequences utilized in the analyses

Access number	Database	Year	Origin	Tree Code
NoV GII.3				
HM072045	GenBank	1975	US	US 1995
HM072046	GenBank	1976	US	US 1976
HM072042	GenBank	1979	US	US 1979
DQ379713	GenBank	1983	Australia	US 1983
AY030312	GenBank	1987	US	US 1987a
AY030313	GenBank	1987	US	US 1987b
HM072044	GenBank	1988	US	US 1988a
HM072043	GenBank	1988	US	US 1988b
U22498	GenBank	1989	Mexico	MEX 1989
L23830	GenBank	1989	Japan	JPN 1989
HM072041	GenBank	1990	US	US 1990
HM072040	GenBank	1991	US	US 1991
U02030	GenBank	1991	Canada	CA 1991
AF414411	GenBank	1993	US	US 1993
AJ277617	GenBank	1993	UK	UK 1993b
AF414413	GenBank	1994	US	US 1994a
AF414412	GenBank	1994	US	US 1994b
AF414414	GenBank	1994	US	US 1994c
U46039	GenBank	1994	New Zealand	NZ 1994
AJ277611	GenBank	1995	UK	UK 1995a
AF414415	GenBank	1995	US	US 1995
AF190817	GenBank	1995	Argentina	ARG 1995
AB195226	GenBank	1997	Japan	JPN 1997a
AB039781	GenBank	1997	Japan	JPN 1997b
AB067542	GenBank	1998	Japan	JPN 1998
AY588132	GenBank	1998	Australia	AUS 1998
DQ093062	GenBank	2000	Japan	JPN 2000a
AB190457	GenBank	2000	Japan	JPN 2000b
AY247431	GenBank	2000	Sweden	SWE 2000
AY247439	GenBank	2001	Sweden	SWE 2001a
AY247441	GenBank	2001	Sweden	SWE 2001b
AF427111	GenBank	2001	Germany	GER 2001a
AF427112	GenBank	2001	Germany	GER 2001b

Access number	Database	Year	Origin	Tree Code
AF539440	GenBank	2001	Germany	GER 2001c
AY845056	GenBank	2002	Australia	AUS 2002
AY652979	GenBank	2003	US	US 2003
AB242256	GenBank	2003	Japan	JPN 2003
AB242258	GenBank	2004	Russia	RU 2004
AB242257	GenBank	2004	Japan	JPN 2004
AB365435	GenBank	2004	US	US 2004
EU850823	GenBank	2005	China	CHN 2005a
EU850825	GenBank	2005	China	CHN 2005b
AB385634	GenBank	2005	Norwey	NTH 2005
AB385641	GenBank	2006	Norwey	NTH 2006
GU339331	GenBank	2007	Mexico	MEX 2007
GU252722	GenBank	2007	UK	UK 2007a
JF262608	GenBank	2008	Vietnam	VNM 2008a
JF262611	GenBank	2008	Vietnam	VNM 2008b
JN602259	GenBank	2008	Australia	AU 2008
JQ934819	GenBank	2010	China	CHN 2010
AJ277617	Noronet	1993	UK	UK 1993b
HCA277611	Noronet	1995	UK	UK 1995b
2123723	Noronet	2001	UK	UK 2001a
2149551	Noronet	2001	UK	UK 2001b
2114494	Noronet	2001	UK	UK 2001c
2125602	Noronet	2001	UK	UK 2001d
Bordeauxs91	Noronet	2001	France	FR 2001
2225876	Noronet	2002	UK	UK 2002a
2212389	Noronet	2002	UK	UK 2002b
2230887	Noronet	2002	UK	UK 2002c
2203728	Noronet	2002	UK	UK 2002d
TH0103b	Noronet	2002	France	FR 2002
VI0104	Noronet	2004	France	FR 2004
BA01	Noronet	2005	Italy	ITA 2005
EVU2284	Noronet	2005	UK	UK 2005a
EVU2592	Noronet	2005	UK	UK 2005b
EVU2296b	Noronet	2005	UK	UK 2005c
HO62700128	Noronet	2006	UK	UK 2006
44LE0506	Noronet	2006	France	FR 2006a
59LI0906	Noronet	2006	France	FR 2006b

Table 2. (Continued)

Access number	Database	Year	Origin	Tree Code
HO70840278	Noronet	2007	UK	UK 2007b
57ME03072	Noronet	2007	France	FR 2007
44NA0208	Noronet	2008	UK	UK 2008a
88EP0308	Noronet	2008	UK	UK 200b
ITMI2009	Noronet	2009	Italy	ITA 2009
115998	Noronet	2009	UK	UK 2009
63CO1211	Noronet	2011	France	FR 2011
NoV GII.7				
AJ277608	GenBank	1990	UK	UK 1990
JN602289	GenBank	2002	Australia	AUS 2002
HM055936	GenBank	2006	Nicaragua	NIC 2006a
HM055934	GenBank	2006	Nicaragua	NIC 2006b
HM055930	GenBank	2006	Nicaragua	NIC 2006c
HM055933	GenBank	2006	Nicaragua	NIC 2006d
HM055931	GenBank	2006	Nicaragua	NIC 2006e
EU735553	GenBank	2006	UK	UK 2006a
GQ849129	GenBank	2007	Australia	AUS 2007
JN602301	GenBank	2008	Australia	AUS 2008a
GQ849130	GenBank	2008	Australia	AUS 2008b
JF896184	GenBank	2008	Korea	KOR 2008a
JF896171	GenBank	2008	Korea	KOR 2008b
HM633199	GenBank	2008	Japan	JPN 2008
GU339333	GenBank	2008	Mexico	MEX 2008a
GU339334	GenBank	2008	Mexico	MEX 2008b
HQ201679	GenBank	2009	South Africa	ZAF 2009
JN596789	GenBank	2009	China	CHN 2009
JN191378	GenBank	2010	South Africa	ZAF 2010
VI0403	Noronet	2003	France	FR 2003
91ES1105	Noronet	2005	France	FR 2005
EVU2852	Noronet	2006	UK	UK 2006b
EVU3260	Noronet	2006	UK	UK 2006c
UD08ISS2	Noronet	2008	Italy	ITA 2008
RCC2710	Noronet	2010	Spain	SP 2010
67WI0111	Noronet	2011	France	FR 2011
NoV GII.13				
AY113106	GenBank	1998	US	US 1998

Access number	Database	Year	Origin	Tree Code
FJ986385	GenBank	2004	Australia	AUS 2004
GU017900	GenBank	2007	Japan	JPN 2007
JF262606	GenBank	2008	Vietnam	VNM 2008a
JF262607	GenBank	2008	Vietnam	VNM 2008b
JF262605	GenBank	2008	Vietnam	VNM 2008c
JN596797	GenBank	2008	China	CHN 2008
JF896161	GenBank	2008	Korea	KOR 2008
GU969058	GenBank	2008	Japan	JPN 2008a
GU017908	GenBank	2008	Japan	JPN 2008b
GU017906	GenBank	2008	Japan	JPN 2008c
GU017904	GenBank	2008	Japan	JPN 2008d
GU017901	GenBank	2008	Japan	JPN 2008e
GU017907	GenBank	2008	Japan	JPN 2008f
GU017905	GenBank	2008	Japan	JPN 2008g
GU017902	GenBank	2008	Japan	JPN 2008h
GU969059	GenBank	2008	Japan	JPN 2008i
JN602314	GenBank	2008	Australia	AUS 2008
HQ201664	GenBank	2008	South Africa	ZAF 2008c
HQ201662	GenBank	2008	South Africa	ZAF 2008b
HQ201659	GenBank	2008	South Africa	ZAF 2008a
JN596796	GenBank	2009	China	CHN 2009
JF909054	GenBank	2009	Ethiopaa	ETH 2009a
JF909055	GenBank	2009	Ethiopia	ETH 2009b
JN191379	GenBank	2010	South Africa	ZAF 2010
JN228271	GenBank	2011	Italy	ITA 2011
59TO0106	Noronet	2005	France	FR 2005
HO62920247	Noronet	2006	UK	UK 2006
35PL0410	Noronet	2010	France	FR 2010
ASPB	Noronet	2010	Spain	SP 2010
NoV GII.14				
AY130761	GenBank	1999	US	US 1999
AJ865520	GenBank	2000	Japan	JPN 2000
AJ865533	GenBank	2001	Japan	JPN 2001
AB078334	GenBank	2002	Japan	JPN 2002
AB354294	GenBank	2003	Japan	JPN 2003
DQ004650	GenBank	2004	Korea	KOR 2004
AJ865582	GenBank	2004	Japan	JPN 2004

Table 2. (Continued)

Access number	Database	Year	Origin	Tree Code
GU017902	GenBank	2008	Japan	JPN 2008
JQ750991	GenBank	2009	China	CHN 2009
2946	Noronet	1999	Norwey	NTH 1999
E2418	Noronet	2005	UK	UK 2005
708	Noronet	2007	Denmark	DK 2007
8698	Noronet	2008	Denmark	DK 2008
1066	Noronet	2008	Sweden	SW 2008a
1090	Noronet	2008	Sweden	SW 2008b
1144	Noronet	2010	Sweden	SW 2010
NoV GII.4				
FJ537135	GenBank	1974	US	US 1974a
FJ537134	GenBank	1974	US	US 1974b
FJ537138	GenBank	1977	US	US 1977
AF414425	GenBank	1995	US	US 1995
JQ965810	GenBank	2003	US	US 2003a
JQ320073	GenBank	2003	US	US 2003b
HQ693090	GenBank	2006	Thailand	THA 2006
JN400609	GenBank	2006	Taiwan	TW 2006a
JN400605	GenBank	2006	Taiwan	TW 2006b
JN899246	GenBank	2006	US	US 2006
JQ182705	GenBank	2006	Mexico	MEX 2006
JQ622197	GenBank	2007	Korea	KOR 2007
JN400613	GenBank	2007	Taiwan	TW 2007a
JN400615	GenBank	2007	Taiwan	TW 2007b
JQ613512	GenBank	2007	Australia	AU 2007
HM635193	GenBank	2008	Korea	KOR 2008a
JN400616	GenBank	2008	Taiwan	TW 2008
HM635181	GenBank	2008	Korea	KOR 2008b
HM635191	GenBank	2008	Korea	KOR 2008c
GQ850882	GenBank	2008	US	US 2008
HM635195	GenBank	2009	Korea	KOR 2009a
JN400618	GenBank	2009	Taiwan	TW 2009a
HM635185	GenBank	2009	Korea	KOR 2009b
HM635183	GenBank	2009	Korea	KOR 2009c
HM635099	GenBank	2009	Korea	KOR 2009d
HM635100	GenBank	2009	Korea	KOR 2009e
JN400617	GenBank	2009	Taiwan	TW 2009b

Access number	Database	Year	Origin	Tree Code
JQ613537	GenBank	2009	Auatralia	AU 2009a
JQ613570	GenBank	2009	Australia	AU 2009b
JF802503	GenBank	2009	Cameroon	CAM 2009
HM535612	GenBank	2009	Taiwan	TW 2009
JQ613527	GenBank	2009	Australia	AU 2009
JQ613552.2	GenBank	2010	Australia	AU 2010a
JN400626	GenBank	2010	Taiwan	TW 2010
HM635189	GenBank	2010	Korea	KOR 2010a
HM635187	GenBank	2010	Korea	KOR 2010b
JQ934814	GenBank	2010	China	CHN 2010
JQ613573	GenBank	2010	Australia	AU 2010b
JQ613571	GenBank	2010	Australia	AU 2010c
JQ613563	GenBank	2010	Australia	AU 2010d
JQ613559	GenBank	2010	Australia	AU 2010e
JQ613553	GenBank	2010	Australia	AU 2010f
JQ613547	GenBank	2010	Australia	AU 2010g
JQ613572	GenBank	2010	Australia	AU 2010h
JQ613558	GenBank	2010	Australia	AU 2010i
JX459907	GenBank	2012	Australia	AU 2012a
JX459908	GenBank	2012	Australia	AU 2012b
JX629458	GenBank	2012	China	CHN 2012
2140311	Noronet	2001	UK	UK 2001
2212309	Noronet	2002	UK	UK 2002
2317184	Noronet	2003	UK	UK 2003
FU0804	Noronet	2004	France	FR 2004
29PA0305	Noronet	2004	France	FR 2005
EVU2364	Noronet	2005	UK	UK 2005a
EVU2494	Noronet	2005	UK	UK 2005b
Lincoln	Noronet	2006	UK	UK 2006a
HO62380227	Noronet	2006	UK	UK 2006b
EVU3513U	Noronet	2006	UK	UK 2006c
71CH0206	Noronet	2006	France	FR 2006a
04BA0407	Noronet	2007	France	FR 2006b
97RE0907	Noronet	2007	France	FR 2006c
MLT1107	Noronet	2007	Italy	ITA 2007
23BO0608	Noronet	2008	France	FR 2008
BZ08ISS3	Noronet	2008	Italy	ITA 2008a

Table 2. (Continued)

Access number	Database	Year	Origin	Tree Code
TS08ISS1	Noronet	2008	Italy	ITA 2008b
ITAB	Noronet	2009	Italy	ITA 2009
63CL1209	Noronet	2009	France	FR 2009a
66PE0109	Noronet	2009	France	FR 2009b
UTE01	Noronet	2010	Spain	SP 2010a
TAR02	Noronet	2010	Spain	SP 2010b
TARGII.4	Noronet	2010	Spain	SP 2010c
UVEVV	Noronet	2010	Spain	SP 2010d
UVEVV	Noronet	2010	Spain	SP 2010e
ASPB	Noronet	2010	Spain	SP 2010f
GIR56	Noronet	2010	Spain	SP 2010g
RCC15	Noronet	2010	Spain	SP 2010h
MI118074Z	Noronet	2010	UK	UK 2010

Time-Measured Phylodynamic Analysis

Evolutionary dynamics including both the rate of nucleotide substitution per site and the time to the most recent common ancestor (TMRCA) for each dataset were estimated using a Bayesian Markov Chain Monte Carlo (MCMC) approach available in the BEAST package v1.7.4 (Drummond and Rambaut, 2007). In each case, both strict and relaxed (uncorrelated-clock log-normal, UCLN) molecular clocks and a substitution model employing the general time reversible (GTR + Γ_4 + I) substitution matrix were used. The Bayesian Skyline population coalescent prior was employed in all cases, since it is the best descriptor for organisms showing cyclic/annual behaviour (Drummond et al., 2005). MCMC chains were run for sufficient time to achieve convergence (assessed using the TRACER program [http://tree.bio.ed.ac.uk/software/tracer/]), with uncertainty in parameter estimates reflected in values of 95% highest posterior density (HPD). Finally BEAST program was used to compute the maximum clade credibility (MCC) tree from all the plausible trees using the TreeAnnotator program, with the first 10% trees removed as burn-in.

Table 3. Nucleotide substitution rates and divergence items over a time period for partial capsid gene of NoV, GII.4, GII.14, GII.3, GII.7

Genotype	Strict molecular clock			Relaxed molecular clock (UCLN)		
	Nucleotide substitution rate[a] (substitutions/site/yr) (95% HPD)[b]	TMRCA[c] (95% HPD) by:		Nucleotide substitution rate[a] (substitutions/site/yr) (95% HPD)[b]	TMRCA[c] (95% HPD) by:	
		Years	Date (range)		Years	Date (range)
GII.4	$7.1x10^{-3}$ ($5.2x10^{-3}$-$9.3x10^{-3}$)	38.6 (37.4-40.2)	1972 (1971-1974)	$1.6x10^{-2}$ ($9.3x10^{-3}$-$2.5x10^{-2}$)	38.7 (37.1-41.5)	1972 (1970-1974)
GII.14	$3.7x10^{-3}$ ($1.6x10^{-3}$-$6x10^{-3}$)	88.3 (41.8-150.5)	1923 (1861-1970)	$1.4x10^{-2}$($8.1x10^{-3}$-$2x10^{-2}$)	20.1 (12.2-33.9)	1991 (1977-1999)
GII.3	$3.7x10^{-6}$($2.5x10^{-3}$-$4.8x10^{-3}$)	37.1 (35.1-40.9)	1974 (1970-1976)	$4.2x10^{-3}$ ($3x10^{-3}$-$5.6x10^{-3}$)	38.2 (36.1-42.6)	1973 (1969-1975)
GII.7	$4.6x10^{-3}$ ($2.6x10^{-3}$-$6.6x10^{-3}$)	21.8 (21.3-23.2)	1989 (1988-1990)	$4.1x10^{-3}$ ($2.2x10^{-3}$-$6.3x10^{-3}$)	22.3 (21.3-24.7)	1989 (1986-1990)
GII.13	$3.28x10^{-3}$ ($1.8x10^{-3}$-$4.9x10^{-3}$)	69.5 (38-107.9)	1942 (1904-1973)	$8.5x10^{-3}$ ($2.4x10^{-3}$-$1.4x10^{-2}$)	21.2 (13.7-45.7)	1990 (1965-1997)
All GII				$5,12x10^{-3}$ ($3.8x10^{-3}$-$6.4x10^{-2}$)	88.1 (49.4-143)	1923 (1868-1962)

[a] mean evolutionary rate for two individuals determination assuming two different molecular clocks

[b] 95% HPD, 95% highest posterior density

[c] TMRCA, time of most recent common ancestor

RESULTS

Rates of Evolutionary Changes and Times of Most Recent Common Ancestry

For each analysed NoV GII dataset, the rate of evolutionary change, expressed as nucleotide substitution per site per year, was inferred. Bayesian MCMC analysis revealed high average rates of nucleotide substitution in NoV GII.4. Thus, assuming a relaxed molecular clock the mean evolutionary rate was 1.6 x 10^{-2} substitution/site/year (95% HPD, 9.3 x 10^{-3}-2.5 x 10^{-2}), whereas assuming a strict-clock model the equivalent mean rate was 7.1 x 10^{-3} substitution/site/year (95% HPD, 5.2 x 10^{-3}-9.3 x 10^{-3}). NoV GII.14 also showed a high mean evolutionary rate compared with the other analysed genotypes under the assumption of a relaxed-clock model, with a value of 1.4 x 10^{-2} substitution/site/year (95% HPD, 8.1 x 10^{-3}-2 x 10^{-2} substitution/site/year). Nonetheless, applying a strict-clock model, mean evolutionary rate showed a lower value of 3.7 x 10^{-3} substitution/site/year (95% HPD, 1.6 x 10^{-3}-6 x 10^{-3} substitution/site/year). This value is similar to those obtained for genotypes GII.7 and GII.13 under relaxed and strict-clock conditions, and for GII.3 applying the relaxed molecular clock (Table 3).

Mean TMRCA estimated under a relaxed molecular clock for each NoV genotype datasets, corresponding to the divergence time of the different linages, showed that NoV GII.14, GII.7 and GII.13 lineages diverged 20 years ago whereas both NoV GII.4 and GII.3 lineages diverged 38 years ago. Results obtained from Bayesian coalescent analyses under both strict or relaxed molecular clock assumption are shown in Table 3.

The coefficient of variation (CoV) obtained under the relaxed-clock model reflects how clock-like is the estimation of substitution rate in the different genotypes. This statistic resumes how the evolutionary rate varies between lineages. In the case of NoV GII.4, GII.14 and GII.13, the high mean value of CoV suggests a good non-clock-like behaviour and is hence indicative of large-scale rate variation among viral lineages. Higher marginal likelihood values in the case of relaxed-clock model of these three genotypes compared with those obtained from strict-clock model analysis, also supported the choice of the relaxed one. In the case of GII.3 and GII.7, although mean values of CoV were lower than those obtained for the other genotypes, also a non clock-like behaviour was shown, being this supported by higher marginal likelihood values compared with analysis assuming strict molecular clock (Table 4).

Taking into account the results obtained for each genotype, an analysis including the five genotypes together was performed under the assumption of a relaxed molecular clock. The obtained results revealed that the global mean evolutionary rate was 5.12 x 10^{-3} substitution/site/year (95% HPD, 3.8 x 10^{-3}-6.4 x 10^{-3}). The TMRCA shared by the analysed genotypes was dated 88 years ago (95% HPD 49-143 years) (Table 3). The mean value of CoV was 2.1, a high value that supports the non-clock-like behaviour of NoV genotypes (Table 4).

Phylogenetic Relationship of NoV Partial Capsid Gene Sequences

MCC trees were inferred for each genotype on the basis of evolutionary analysis to determine a time-measured phylogenetic relationship of partial capsid gene sequences. The MCC tree for NoV GII.4 dataset (37-years period) (Figure 1) included sequences belonging to different variants associated with global epidemics of acute gastroenteritis from 1996 to present. These variants were US 95/96 in 1996, Farmington Hills in 2002, Hunter in 2004, 2006b in 2007 and 2008, New Orleans virus from 2009 to 2012, and Sydney 2012. Another three variants were also included in the analysis, named 2006a, Osaka 2007 and Apeldoorn 2008, which were associated with localized epidemics rather than global pandemics (Eden et al., 2013). Phylogenetic relations based on partial ORF2 sequences showed that the 2006b and Osaka 2007 clusters were grouped together in the same branch (named A), whereas 2006a, Apeldoorn 2008, New Orleans 2009 and Sydney 2012 clusters were located in a second main branch (named B). These two main branches shared a TMRCA with a mean date of divergence of 6.5 years (year 2006) (95%HPD, 6-7,5 years).

There were also other four well differentiated clusters, corresponding to the sequences from the CHDC (Children's Hospital National Medical Center; Washington DC) obtained in the 1970s, the US 95/96 sequences, the Farmington Hill 2002 sequences and the Hunter 2004 sequences. Strains from NW Spain belonging to variant 2006b diverged from a TMRCA with a mean date of divergence of 3 years (year 2009) (95% HPD, 1,5-3 years) together with one strain detected in Korea and one strain detected in France. Spanish sequences obtained in the present study and belonging to 2009 variant, 2012 variant and unassigned variants were located in the second main branch B. Most of these strains evolved together with other Spanish strains associated with several outbreaks in different locations. Reference strains of the new emerging 2012 variant (named AU2012a, AU2012b and CHN2012 in the tree) shared a common ancestor with five of the Spanish strains assigned to these variant, dated in 2011 (95% HPD, 0.5-1 years).

Table 4. CoV estimation and Marginal likelihood for both strict and relaxed models as a measure of the quality of the analysis

	Strict molecular clock		Relaxed molecular clock (UCLN)	
Genotype	CoV (95% HPD)	Log P(M/D)[a] (S.E.)[b]	CoV (95% HPD)	Log P(M/D)
GII.4	NA[c]	-1931.2 (+/-24)	1.7	-1873 (+/- 0.5)*
GII.14	NA	-2075.1 (+/- 0.2)	2.57 (1.4-4)	-1851.9 (+/- 0.2)*
GII.3	NA	-1716.4 (+/- 0.1)	0.85 (0.5-1.2)	-1716.2 (+/- 0.1)*
GII.7	NA	-765 (+/- 0.1)	0.5 (0.2-1.2)	-745.2 (+/-0.1)*
GII.13	NA	-1078 (+/- 0.1)	2.3 (1.1-3.5)	-1063 (+/- 0.1)*
All GII genotypes			2.1 (1.6-2.6)	- 4941 (+/- 0.3)

[a] Marginal likelihood reported for each dataset considered under two different molecular clocks. Chosen molecular clock is marked (*)

[b] Smoothed estimate using 1000 bootstrap replicates

[c] NA, not applicable

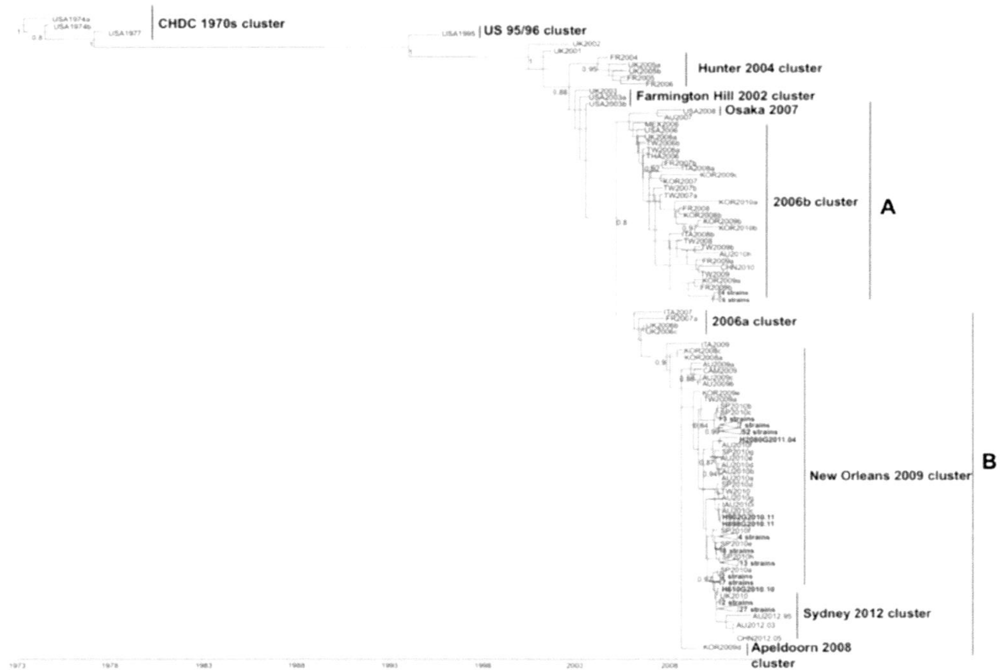

Figure 1. Maximum clade credibility tree of 213 NoV GII.4 sequences (37-year period). Strains from NW of Spain are indicated and represented in the collapsed branches or remarked in black. All tip time on the tree corresponding to year of sampling (as reflected on the timescale on the x axis). The shaded bars are each node represent the 95% HPD values for node height (age), while posterior probability values (a measurement of clustering support) are shown for the key nodes. The tree was rooted on the assumption of a relaxed molecular clock.

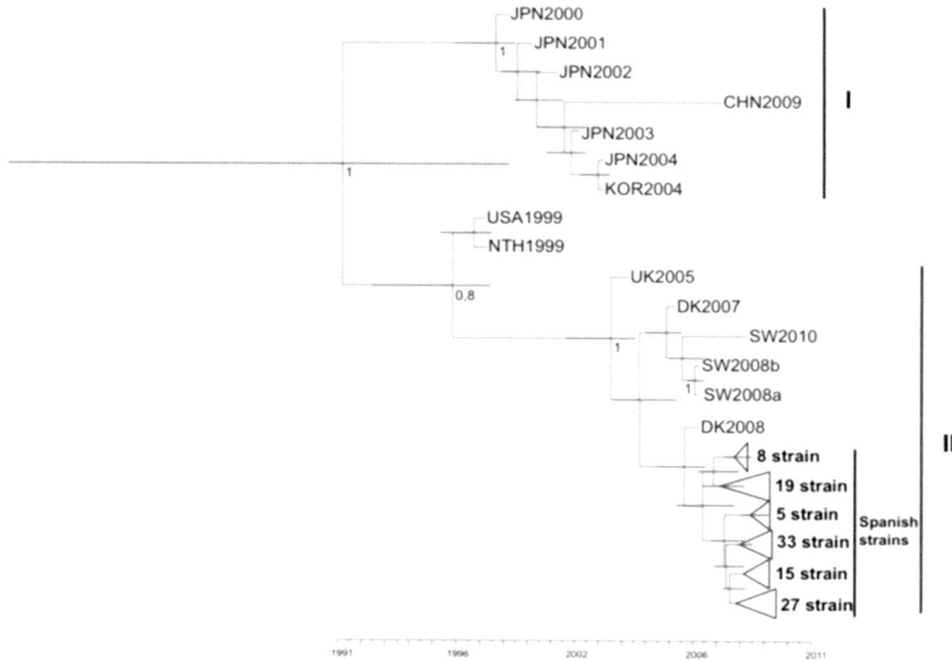

Figure 2. Maximum clade credibility tree of 123 NoV GII.14 sequences (12-year period). Strains from NW of Spain are indicated and represented in the collapsed branches. All tip time on the tree corresponding to year of sampling (as reflected on the timescale on the x axis). The shaded bars are each node represent the 95% HPD values for node height (age), while posterior probability values (a measurement of clustering support) are shown for the key nodes. The tree was rooted on the assumption of a relaxed molecular clock.

The MCC tree for NoV GII.14 dataset (12-years period) (Figure 2) showed two well-differentiated clusters. Cluster I comprised strains from Japan, China and Korea dated from 2000 to 2009. In cluster II were located strains from different European countries dated from 2005 to 2010. These European linages diverged from a common ancestor with a mean date of 7.5 years (year 2003) (95% HPD, 7.5-9.5 years). Also, these lineages seem to evolve from a TMRCA shared with US and Netherland strains dated of 15 years (year 1996) (95% HPD, 1-2 years). Spanish strains from the present work diverged from a common ancestor together with a Danish strain, dated from 5.5 years (95% HPD, 4.5-6.5 years), and were separated into two groups, with a mean diverged date of 4.5 years (95% HPD, 3-5.5 years).

The MCC tree obtained from the NoV GII.3 (36-year period) (Figure 3) showed that Spanish strains established a well-differentiated cluster that shared a common ancestor with a strain from US isolated in 1990. This TMRCA had a mean divergence date of 23 years (95% HPD, 21-27 years).

In the case of NoV GII.7 the analysis (21-year period) rendered a MCC tree (Figure 4) showing two main branches. In one of them (named 1) were located strains from United Kingdom and Nicaragua, isolated all of them in 2006. The strains from United Kingdom shared also a common ancestor with strains from China and Japan. In the second branch (named 2) were located the strains detected from 2005 to 2010 in other different parts of the world. These two main clusters share a TMRCA with a divergence date of 10.5 years (95% HPD, 0.5-3 years). Most of Spanish strains were distributed in 3 clusters, sharing a common ancestor with a strain from France, with a divergence date of 3 years (95% HPD, 2-4.5 years). The MCCR tree obtained from NoV GII.13 dataset (13-years period) (Figure 5) showed that all the strains from 2004 to 2011, regardless their country of isolation, shared the same TMRCA with a divergence date of 11.5 years (year 1999) (95% HPD, 0.5-3 years). Most Spanish strains were grouped in the same branch together with a strain isolated from Italy, also sharing with a strain from France a common ancestor dated in 2010 (95% HPD, 1.5-3.5 years).

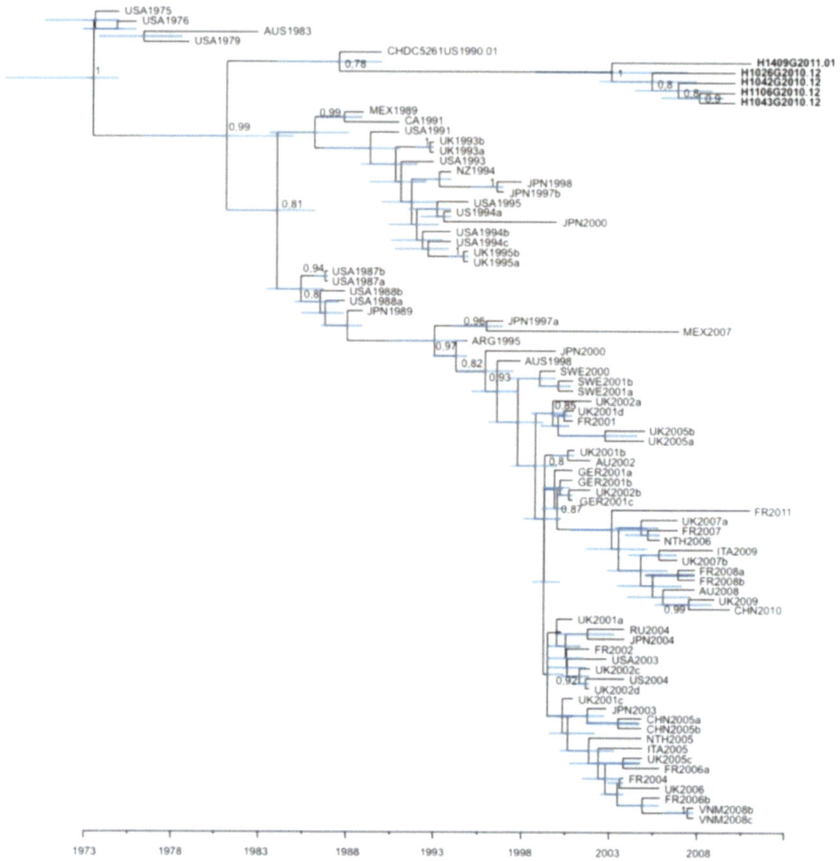

Figure 3. Maximum clade credibility tree of 82 NoV GII.3 sequences (36-year period). Strains from NW of Spain are remarked in black. All tip time on the tree corresponding to year of sampling (as reflected on the timescale on the x axis). The shaded bars at each node represent the 95% HPD values for node height (age), while posterior probability values (a measurement of clustering support) are shown for key nodes. The tree was rooted on the assumption of a relaxed molecular clock.

The MCC tree including the five NoV GII genotypes revealed that they diverged from a TMRCA dated 88 years ago (95% HPD, 34-40 years). In one of the branches were located the lineages corresponding to NoV GII.4 and GII.14. The TMRCA shared by NoV GII.4 and GII.14 dated 52 years ago (95% HPD, 11-17 years). In the other branch were located NoV GII.3, GII.13 and GII.7. NoV GII.3 diverged 50 years ago (95% HPD, 10-25 years). The most recent genotypes were NoV GII.13 and GII.7, sharing both a common ancestor with a divergence date of 31 years (95% HPD, 10-15 years) (Figure 6).

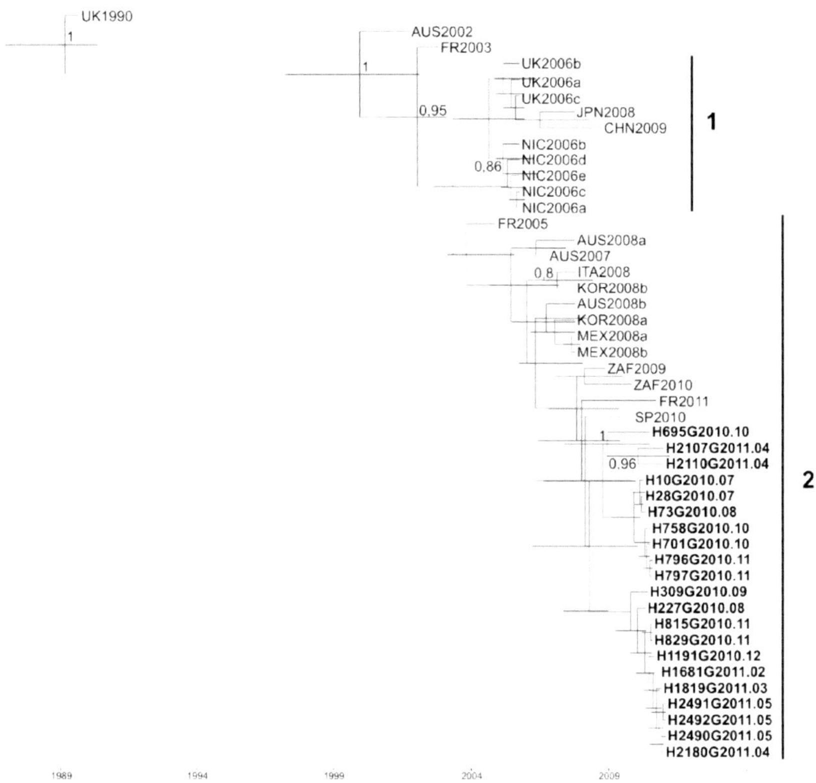

Figure 4. Maximum clade credibility tree of 47 NoV GII.7 sequences (21-year period). Strains from NW of Spain are remarked in black. All tip time on the tree corresponding to year of sampling (as reflected on the timescale on the x axis). The shaded bars at each node represent the 95% HPD values for node height (age), while posterior probability values (a measurement of clustering support) are shown for key nodes. The tree was rooted on the assumption of a relaxed molecular clock.

Evolutionary Dynamics Inference

The Bayesian Skyline Plot (BSP) model was used to infer the population history of different NoV genotypes. The BSP shows the relative genetic diversity of a population over time, where N_e is the effective population size and τ is the generation. $N_e\tau$ reflects the number of effective infection (replication) established by the virus. These relative genetic diversity plots can help to illustrate a demographical history by showing increases and decreases of genetic diversity within a defined time frame. The Lineage Through Time Plot (LTTP) was also used to get a sense of the overall pattern of diversification expressed as the number of lineages that had any successful descendant against time. NoV GII.4 BSP showed a slight increase of $N_e\tau$ value in 2002, getting the highest values during 2006-2009, with a peak also in 2010 (Figure 7a).

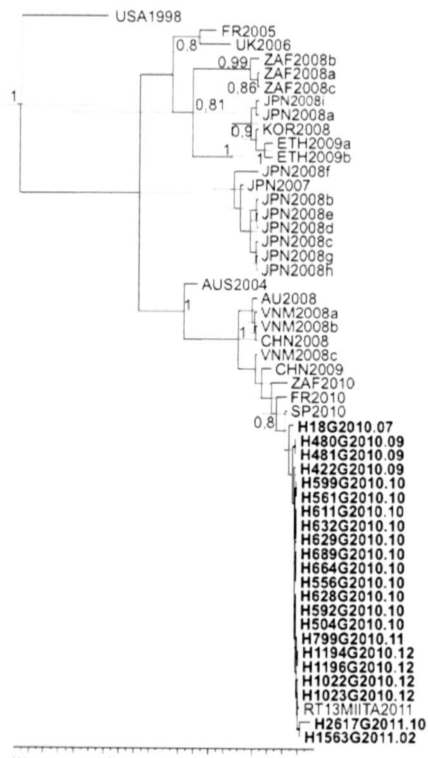

Figure 5. Maximum clade credibility tree of 52 NoV GII.13 sequences (13-year period). Strains from NW of Spain are remarked in black. All tip time on the tree corresponding to year of sampling (as reflected on the timescale on the x axis). The shaded bars at each node represent the 95% HPD values for node height (age), while posterior probability values (a measurement of clustering support) are shown for key nodes. The tree was rooted on the assumption of a relaxed molecular clock.

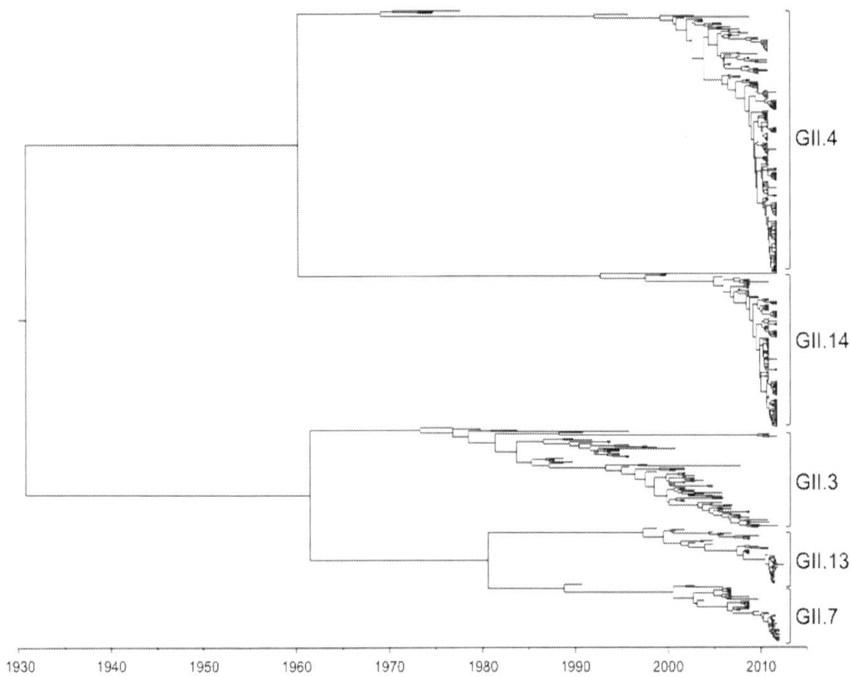

Figure 6. Maximum clade credibility tree of NoV GII.4, GII.14, GII.3; GII.7 and GII.13 sequences. The shaded bars at each node represent the 95% HPD values for node height (age), while posterior probability values (a measurement of clustering support) are shown for key nodes. The tree was rooted on the assumption of a relaxed molecular clock.

These results are in agreement with an increment in the appearance of lineages through time (Figure 7b). The LTTP seems to reflect the well-known appearance of 1995, 2002, 2004, 2006a and b, and 2009 variants. The 2002 variant, which had a strong impact in the population according to surveillance data, did not coincide with a pronounced upsurge in the BSP, whereas the appearance of 2009 variant seems to coincide with a peak in $N_e\tau$ value.

The analysis of BSP and LTTP for NoV GII.14 (Figure 7c,d) showed an increase in $N_e\tau$ values during the year 2009, with peaks in 2010 and 2011, indicating an important increment in the number of lineages appeared during these years. It is important to highlight that BSP for NoV GII.14 appears to be inversely correlated with GII.4 BSP. The peak in NoV

GII.14 BSP at the beginning of 2010 coincide with a notable decrease in NoV GII.4 $N_e\tau$ value, whereas the peak in $N_e\tau$ values for GII.4 during that year (and that seems to coincide with the emerging of 2009 variant) are correlated with a decrement in $N_e\tau$ values for GII.14 during the same time-period.

The BSP obtained for NoV GII.3 (Figure 7e) showed a high increase of $N_e\tau$ during the year 2000, with an increase in effective population size that remains constant until 2011. NoV GII.7 showed a BSP (Figure 7g) with an evolutionary dynamic similar to that observed for NoV GII.4, with the highest peak of activity during 2006-2009. In contrast to the rest of the analysed genotypes, GII.13 BSP (Figure 7i) showed a decrease in $N_e\tau$ value during a long time-period from 2002 to 2010. In these three genotypes, especially in the case of NoV GII.3, nucleotide substitution rates are lower than the ones obtained for GII.4 and GII.14. The LTTP (Figure 7f, h, j) in the three cases showed also a low pattern of diversification expressed as the number of lineages within a defined time frame.

Discussion

NoV are important causative agents of acute gastroenteritis in both children and adults worldwide, but the mechanisms responsible for the emergence of successful predominant strains are poorly understood (Hoehne and Schreier, 2006; Iturriza-Gomara et al., 2008). The majority of studies on the evolution of NoV have been focused broadly in GII.4, because its high prevalence in the population (Lindesmith et al., 2008; Bok et al., 2009; Victoria et al., 2009; Siebenga et al., 2010). During the years 2010-2011 there was a co-circulation of different GII genotypes in NW of Spain, similarly to other Spanish regions, being GII.4 the predominant genotype (Manso and Romalde, 2013; Arana et al., 2014; Carmona-Vicente et al., 2015). However, a high detection rate was also observed for NoV GII.14 in the analysed population during that period (Manso and Romalde, 2013). The goal of this work was to compare the population dynamics and the evolutionary rates of the different co-circulating genotypes and also determine the TRMCA of the NoV strains detected in NW of Spain belonging to different GII genotypes. The understanding of NoV evolutionary dynamics would have significant epidemiological importance, predicting the emergence of new strains and in the development of vaccines and the establishment of vaccination programs.

A key component in establishing evolutionary mechanism of NoV is the knowledge of the rate at which NoV generate genetic diversity that leads to fixed mutations in the viral population. Most RNA viruses evolve at a rate of 10^{-3} nucleotide substitution per site per year (Duffy et al., 2008). The analysis of partial capsid gene sequence showed that NoV GII.3, GII.7 and GII.13 evolve at a comparable rate over the established time-periods. Similar results were previously found by other authors for NoV GII.3 using the complete VP1 gene (Boon et al., 2011). Obtained rates for NoV GII.4 and GII.14 were higher (10^{-2}), similar to those reported in other studies for NoV GII.4 analysing both complete or partial capsid gene region (Victoria et al., 2009).

The high evolutionary rates might have allowed the NoV to generate replacement clusters and constantly evade the immune response, within the same genotype (Siebenga et al., 2007). NoV GII.4, which has a high evolutionary rate, is the most common genotype detected in human populations (Lopman et al., 2004; Kroneman et al., 2008; Siebenga et al., 2009; Bull et al., 2010). The MCC tree obtained in this work for NoV GII.4 seams to reflect the continuous linage turnover through time and how new variants evolves quickly from previous existing ones. Also, the results obtained in basis of partial ORF2 sequences are in agreement with previous finding revealing that the ORF2 of New Orleans 2009 variants derived from Apeldoorn 2008 variant, and Sydney 2012 variant derived both from New Orleans 2009 and Apeldoorn 2008 variants (Eden et al., 2013). Furthermore, the ORF2 of these three variants (New Orleans 2009, Apeldoorn 2008, and Sydney 2012) have a distinct non 2006b-like origin (Eden et al., 2013).

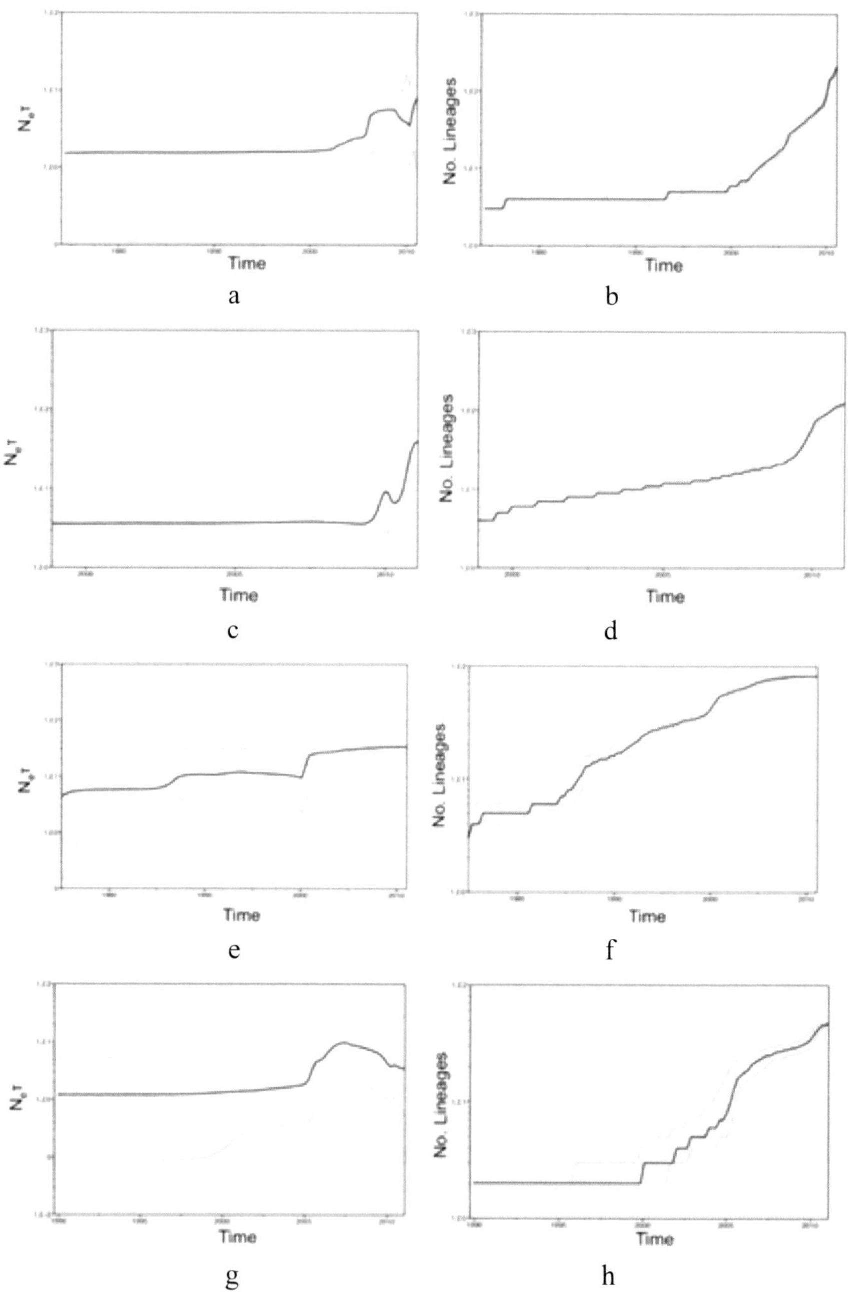

Figure 7. (Continued)

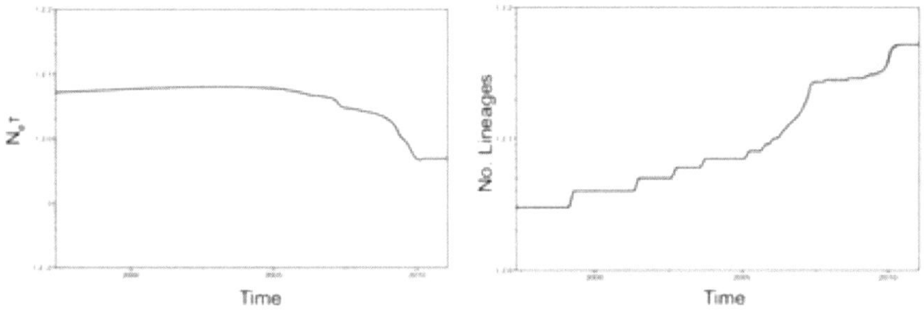

i j

Figure 7. Bayesian Skyline Plots (BSP) and Lineage through Time Plots (LTTP) for different NoV genotypes. BSP illustrate the relative genetic diversity through time whereas LTTP illustrate the overall pattern of diversification expressed as the number of lineages that had any living descendant against time. In each case the black line represent the median posterior value and the grey lines the 95% highest posterior density (HPD) intervals: a) GII.4 BSP, b) GII.4 LTTP, c) GII.14 BSP, d) GII.14 LTTP, e) GII.3 BSP, f) GII.3 LTTP, g) GII.7 BSP, h) GII.7 LTTP, i) GII.13 BSP, j) GII.13 LTTP.

The resulting BSP for NoV GII.4 obtained from this work showed a slight increase in $N_e\tau$ during the year 2002, whereas from year 2006 to 2009, the highest values for this parameter were observed, with a new increment in 2010. The increases in $N_e\tau$ values agree with the appearance and expansion of new variants, as shows the LTTP. In the case of variant 2002, it is known that the replacement of US95/96 variant by 2002 variant became in a loss of genetic diversity (Siebenga et al., 2010), so the lack of correlation between the high impact it had in the population and the slight increase in $N_e\tau$ value may be explained by the fact that BSP does not reflect properly bottle-neck events in the population.

A high evolutionary rate for NoV GII.14 was also found in the present study, which could be correlated with the high detection rate of this genotype in the analysed human population from NW of Spain during the two last years due to the emergence and widespread of new genotype variants. This genotype, together with GII.4, are the ones with highest evolutionary rate and are also the most ancient ones, those that diverged first from a common ancestor. The BSP revealed an alternate pattern in the $N_e\tau$ values of these two genotypes.

It has been proposed that lower rate of evolution and a less processive polymerase implies an evolutionary disadvantage of other NoV GII genotypes compared to NoV GII.4 (Bull et al., 2010). An evolutionary dynamic similar to GII.4 was observed for NoV GII.7, but the nucleotide substitution rate was lower for the latter genotype. This results support the previous findings reported by Bull et al. (2010) for this genotype, in which despite having a high degree of replication, the low nucleotide substitution rate limits its adaptation and therefore its incidence. In the case of NoV GII.13, the evolutionary disadvantage compared with other genotypes is due to a lower nucleotide substitution rate and a decrease in the number of effective replications established by the virus ($N_e\tau$) for a long time-period. NoV GII.3 showed an evolutionary pattern with a constant effective population size from year 2000 to 2011. This genotype is described as one of the most common genotype associated with sporadic NoV infections, particularly in children, where it is often identified as the dominant genotype (Dove et al., 2005; Phan et al., 2005; Bok et al., 2009; Barreira et al., 2010; Dey et al., 2010). It is important to note the low detection rate observed in the analysed Spanish population and the fact that Spanish GII.3 strains showed a TMRCA with an average divergence date of 23 years, that means a slow emergence of GII.3 variants in this geographical region.

The epidemiological fitness on NoV GII.4 is a consequence of the ability of the virus to generate genetic diversity, as the NoV pandemic GII.4 strains were associated with an increased replication and mutation rate (20). Results obtained from these analyses revealed that, whereas some genotypes as GII.3, GII.7 and GII.13 showed an evolutionary disadvantage in comparison with GII.4, NoV GII.14 demonstrates a nucleotide substitution rate and evolutionary dynamic similar to NoV GII.4, suggesting the future emergence and widespread epidemics of other genotype variants (Iritani et al., 2008).

In summary, this study provides an insight into the complex evolutionary patterns of different NoV GII genotypes and reports the first estimation of evolutionary rates for GII.7, GII.13 and GII.14. The understanding of evolutionary dynamics of different NoV genotypes can be helpful in the development of suitable vaccine candidates and other preventing measures targeting the most likely strains circulating in a geographical area.

REFERENCES

Ando T, Noel JS, Fankhauser RL. 2000. Genetic classification of "Norwalk-like viruses." *J. Infect. Dis*. 181:336-348.

Arana A, Cilla G, Montes M, Gomariz M, Pérez-Trallero E. 2014. Genotypes, recombinant forms, and variants of Norovirus GII.4 In: Gipuzkoa (Basque Country, Spain), 2009-2012. *PloS ONE* 9:e98875.

Barreira DM, Ferreira MS, Fumian TM, Checon R, de Sadovsky AD, Leite JP, Miagostovich MP, Spano LC. 2010. Viral load and genotypes of noroviruses in symptomatic and asymptomatic children in Southeastern Brazil. *J. Clin. Virol.* 47:60-64.

Bok K, Abente EJ, Realpe-Quintero M, Mitra T, Sosnovtsev SV, Kapikian AZ, Green KY. 2009. Evolutionary dynamics of GII.4 noroviruses over a 34-year period. *J. Virol.* 83:11890-11901.

Boon D, Mahar JE, Abente EJ, Kirkwood CD, Purcell RH, Kapikian AZ, Green KY, Bok K. 2011. Comparative evolution of GII.3 and GII.4 norovirus over a 31-year period. *J. Virol.* 85:8656-8666.

Bull RA, Eden JS, Rawlinson WD, White PA. 2010. Rapid evolution of pandemic noroviruses of the GII.4 lineage. *PLoS Pathog*. 6:e1000831.

Carmona-Vicente N, Fernández-Jiménez M, Ribes JM, Téllez-Castillo CJ, Khodayar-Pardo P, Rodríguez-Diaz J, Buesa J. 2015. Norovirus infections and seroprevalence of genotype GII.$-specific antibodies in a Spanish population. *J. Med. Virol.* 87:675-682.

Chan MC, Sung JJ, Lam RK, Chan PK, Lee NL, Lai RW, Leung WK. 2006. Fecal viral load and norovirus-associated gastroenteritis. *Emerg. Infect. Dis.* 12:1278-1280.

Dey SK, Phathammavong O, Okitsu S, Mizuguchi M, Ohta Y, Ushijima H. 2010. Seasonal pattern and genotype distribution of norovirus infection in Japan. *Pediatr. Infect. Dis. J.* 29:e32-4.

Donaldson EF, Lindesmith LC, Lobue AD, Baric RS. 2008. Norovirus pathogenesis: mechanisms of persistence and immune evasion in human populations. *Immunol. Rev.* 225:190-211.

Dove W, Cunliffe NA, Gondwe JS, Broadhead RL, Molyneux ME, Nakagomi O, Hart CA. 2005. Detection and characterization of human caliciviruses in hospitalized children with acute gastroenteritis in Blantyre, Malawi. *J. Med. Virol.* 77:522-527.

Drummond AJ, Rambaut A. 2007. BEAST: Bayesian evolutionary analysis by sampling trees. BMC Evol Biol 7:214.

Drummond AJ, Rambaut A, Shapiro B, Pybus OG. 2005. Bayesian coalescent inference of past population dynamics from molecular sequences. *Mol. Biol. Evol.* 22:1185-1192.

Duffy S, Shackelton LA, Holmes, EC. 2008. Rates of evolutionary change in viruses: patterns and determinants. *Nat. Rev. Genet.* 9:267-276.

Eden JS, Tanaka MM, Boni MF, Rawlinson WD, White PA. 2013. Recombination within the pandemic norovirus GII.4 lineage. *J. Virol.* 87:6270-6282.

Glass RI, Parashar UD, Estes MK. 2009. Norovirus gastroenteritis. *N. Engl. J. Med.* 361:1776-1785.

Green SM, Dingle KE, Lambden PR, Caul EO, Ashley CR, Clarke IN. 1994. Human enteric Caliciviridae: A new prevalent small roundstructured virus group defined by RNA-dependent RNA polymerase and capsid diversity. *J. Gen. Virol.* 75:1883-1888.

Harris JP, Edmunds WJ, Pebody R, Brown DW, Lopman BA. 2008. Deaths from norovirus among the elderly, England and Wales. *Emerg. Infect. Dis.* 14:1546-1552.

Hoehne M, Schreier, E. 2006. Detection of norovirus genogroup I and II by multiplex real-time RT-PCR using a 3'-minor groove binder-DNA probe. *BMC Infect. Dis.* 6:69.

Hoffmann D, Seebach J, Foley BT, Frösner G, Nadas K, Protzer U, Schätzl HM. 2010. Isolated norovirus GII.7 strain within an extended GII.4 outbreak. *J. Med. Virol.* 82:1058-1064.

Iritani N, Kaida A, Kubo H, Abe N, Goto K, Ogura H, Seto Y. 2010. Molecular epidemiology of noroviruses detected in seasonal outbreaks of acute nonbacterial gastroenteritis in Osaka City, Japan, from 1996-1997 to 2008-2009. *J. Med. Virol.* 82:2097-2105.

Iritani N, Vennema H, Siebenga JJ, Siezen RJ, Renckens B, Seto Y, Kaida A, Koopmans M. 2008. Genetic analysis of the capsid gene of genotype GII.2 noroviruses. *J. Virol.* 82:7336-7345.

Iturriza-Gomara M, Xerry J, Gallimore CI, Dockery C, Gray J. 2008. Evaluation of the Loopamp (loop-mediated isothermal amplification) kit for detecting norovirus RNA in faecal samples. *J. Clin. Virol.* 42:389-393.

Johnson PC, Mathewson JJ, DuPont HL, Greenberg HB. 1990. Multiple-challenge study of host susceptibility to Norwalk gastroenteritis in US adults. *J. Infect. Dis.* 161:18-21.

Koopmans M, Vinjé J, de Wit M, Leenen I, van der Poel W, van Duynhoven Y. 2000. Molecular epidemiology of human enteric caliciviruses in The Netherlands. *J. Infect. Dis.* 18:262-269.

Kroneman A, Verhoef L, Harris J, Vennema H, Duizer E, van Duynhoven Y, Gray J, Iturriza M, Böttiger B, Falkenhorst G, Johnsen C, von Bonsdorff CH, Maunula L, Kuusi M, Pothier P, Gallay A, Schreier E, Höhne M, Koch J, Szücs G, Reuter G, Krisztalovics K, Lynch M, McKeown P, Foley B, Coughlan S, Ruggeri FM, Di Bartolo I, Vainio K, Isakbaeva E, Poljsak-Prijatelj M, Grom AH, Mijovski JZ, Bosch A, Buesa J, Fauquier AS, Hernandéz-Pezzi G, Hedlund KO, Koopmans M. 2008. Analysis of integrated virological and epidemiological reports of norovirus outbreaks collected within the foodborne viruses in Europe network from 1 July 2001 to 30 June 2006. *J. Clin. Microbiol.* 46:2956-2965.

Leon JS, Kingsley DH, Montes JS, Richards GP, Lyon GM, Abdulhafid GM, Fernandez ML, Teunis PF, Flick GJ, Moe CL. 2011. Randomized, double-blinded clinical trial for human norovirus inactivation in oysters by high hydrostatic pressure processing. *Appl. Environ. Microbiol.* 77:5476-5482.

Lewis DC, Hale A, Jiang X, Eglin R, Brown DW. 1997. Epidemiology of Mexico virus, a small round-structured virus in Yorkshire, United Kingdom, between January 1992 and March 1995. *J. Infect. Dis.* 175:951-954.

Lindesmith LC, Donaldson EF, Lobue AD, Cannon JL, Zheng DP, Vinje J, Baric RS. 2008. Mechanisms of GII.4 Norovirus Persistence in Human Populations. *PLoS Med.* 5:e31.

Lopman BA, Reacher MH, Vipond IB, Sarangi J, Brown DW. 2004. Clinical manifestation of norovirus gastroenteritis in health care settings. *Clin. Infect. Dis.* 39:318-324.

Manso CF, Romalde JL. 2013. Role of Norovirus in acute gastroenteritis development in the northwest of Spain during 2010-2011 period. *J. Med. Virol.* 85:2009-2015.

Medici MC, Tummolo F, Martella V, Giammanco GM, De Grazia S, Arcangeletti MC, De Conto F, Chezzi, C, Calderaro A. 2014. Nover recombinant GII.P16_GII.13 and GII.P16_GII.3 norovirus strains in Italy. *Virus Res.* 188:142-145.

Motomura K, Oka T, Yokoyama M, Nakamura H, Mori H, Ode H, Hansman GS, Katayama K, Kanda T, Tanaka T, Takeda N, Sato H, Norovirus Surveillance Group of Japan. 2008. Identification of monomorphic and divergent haplotypes in the 2006-2007 norovirus GII/4 epidemic population by genomewide tracing of evolutionary history. *J. Virol.* 82:11247-11262.

Okada M, Tanaka T, Oseto M, Takeda N, Shinozaki K. 2006. Genetic analysis of noroviruses associated with fatalities in healthcare facilities. *Arch. Virol.* 151:1635-1641.

Patel MM, Hall AJ, Vinjé J, Parashar UD. 2009. Noroviruses: a comprehensive review. *J. Clin. Virol.* 44:1-8.

Phan TG, Nguyen TA, Nishimura S, Nishimura T, Yamamoto A, Okitsu S, Ushijima H. 2005. Etiologic agents of acute gastroenteritis among Japanese infants and children: virus diversity and genetic analysis of sapovirus. *Arch. Virol.* 150:1415-1424.

Rockx B, De Wit M, Vennema H, Vinje J, De Bruin E, Van Duynhoven Y, Koopmans M. 2002. Natural history of human calicivirus infection: a prospective cohort study. *Clin. Infect. Dis.* 35:246-253.

Sabriá A, Pintó RM, Bosch A, Bartolomé R, Cornejo T, Torner N, Martínez A, de Simón M, Domínguez A, Guix S, Catalan Viral Gastroenteritis Study Group. 2014. Molecular and clinical epidemiology of norovirus outbreaks in Spain during the emergence of GII.4 2012 variant. *J. Clin. Virol.* 60:96-104.

Siebenga JJ, Beersma MF, Vennema H, van Biezen P, Hartwig NJ, Koopmans M. 2008. High Prevalence of Prolonged Norovirus Shedding and Illness among Hospitalized Patients: A Model for *In Vivo* Molecular Evolution. *J. Infect. Dis.*, 198: 994-1001.

Siebenga JJ, Lemey P, Kosakovsky Pond SL, Rambaut A, Vennema H, Koopmans M. 2010. Phylodynamic reconstruction reveals norovirus GII.4 epidemic expansions and their molecular determinants. *PLoS Pathog.* 6:e1000884.

Siebenga JJ, Vennema H, Renckens B, de Bruin E, van der Veer B, Siezen RJ, Koopmans M. 2007. Epochal evolution of GGII.4 norovirus capsid proteins from 1995 to 2006. *J. Virol.* 81: 9932-9941.

Siebenga JJ, Vennema H, Zheng DP, Vinjé J, Lee BE, Pang XL, Ho EC, Lim W, Choudekar A, Broor S, Halperin T, Rasool NB, Hewitt J, Greening GE, Jin M, Duan ZJ, Lucero Y, O'Ryan M, Hoehne M, Schreier E, Ratcliff RM, White PA, Iritani N, Reuter G, Koopmans M. 2009. Norovirus illness is a global problem: emergence and spread of norovirus GII.4 variants, 2001-2007. *J. Infect. Dis.* 200:802-812.

Tamura K, Peterson D, Peterson N, Stecher G, Nei M, Kumar S. 2011. MEGA5: Molecular evolutionary genetics analysis using Maximum Likelihood, evolutionary distance, and Maximum Parsimony methods. *Mol. Biol. Evol.* 28:2731-2739.

Teunis PF, Moe CL, Liu P, Miller SE, Lindesmith L, Baric RS, Le Pendu J, Calderon RL. 2008. Norwalk virus: how infectious is it? *J. Med. Virol.* 80:1468-1476.

Tu ET, Bull RA, Greening GE, Hewitt J, Lyon MJ, Marshall JA, McIver CJ, Rawlinson WD, White PA. 2008. Epidemics of gastroenteritis during 2006 were associated with the spread of norovirus GII.4 variants 2006a and 2006b. *Clin. Infect. Dis*. 46:413-420.

Victoria M, Miagostovich MP, Ferreira MS, Vieira CB, Fioretti JM, Leite JP, Colina R, Cristina J. 2009. Bayesian coalescent inference reveals high evolutionary rates and expansion of Norovirus populations. *Infect. Genet. Evol.* 9:927-932.

Wang J, Jiang X, Madore HP, Gray J, Desselberger U, Ando T, Seto Y, Oishi I, Lew JF, Green KY. 1994. Sequence diversity of small, round-structured viruses in theNorwalk virus group. *J. Virol*. 68:5982-5990.

de Wit MA, Koopmans MP, Kortbeek LM, Wannet WJ, Vinje J, van Leusden F, Bartelds AI, van Duynhoven YT. 2001. Sensor, a population-based cohort study on gastroenteritis in the Netherlands: incidence and etiology. *Am. J. Epidemiol*. 154:666-674.

Zheng DP, Ando T, Fankhauser RL, Beard RS, Glass RI, Monroe SS. 2006. Norovirus classification and proposed strain nomenclature. *Virol*. 346:312-323.

In: Noroviruses
Editor: Jesús L. Romalde
ISBN: 978-1-53611-939-8

Chapter 5

CONSIDERATIONS FOR EXPOSURE AND DOSE-RESPONSE ASSESSMENT IN NOROVIRUS QMRA

Nicole A. Van Abel[1,*], Jason Shao[2], Mary E. Schoen[3], John C. Kissel[1] and J. Scott Meschke[1,4]

[1]Department of Environmental and Occupational Health Sciences, University of Washington, Seattle, WA, US
[2]Department of Biostatistics, University of Washington, Seattle, WA, US
[3]Soller Environmental, Berkeley, California, US
[4]Department of Civil and Environmental Engineering, University of Washington, Seattle, WA, US

* Corresponding Author address: Department of Environmental and Occupational Health Sciences, University of Washington 4225 Roosevelt Way NE, Suite 100, Seattle, WA 98195, United States. Email: nvanabel@uw.edu

ABSTRACT

Norovirus is a major public health problem leading to numerous global food and waterborne outbreaks. Quantitative Microbial Risk Assessments (QMRAs) are increasingly being developed for norovirus that include numerous assumptions in the exposure and dose-response assessments because of limited information. Despite recently described tissue culture infectivity methods described for norovirus, the concentration of norovirus is most typically still estimated by molecular methods that determine the presence of pathogens, but do not differentiate between viable and non-viable particles. As a result, the infectivity of norovirus is commonly assumed for the sake of QMRA. Innate susceptibility (or secretor status) and acquired immunity are still not fully understood and vary based on ancestry, biology, and previous exposure. In addition, several proposed dose-response models have now been described for use in QMRAs for norovirus. However, the determination of which model is best is still a matter of contention. Norovirus dose-response models have many considerations, such as secretor status, genogroup, and aggregation, and yield different risk estimates depending on the assumption of aggregated or disaggregated dose. Future research needs include determination of the infectivity of norovirus from exposure matrices as measured by cell culture, a better understanding of norovirus aggregation, more low dose human challenge data across a range of viral genotypes, studies on genogroup differences in immunity mechanisms, and clarification of the dose-response among individuals with and without aquired immunity.

Keywords: exposure assessment, dose response, QMRA

INTRODUCTION

Norovirus is a major problem worldwide and has been recognized as a major cause of sporadic and epidemic gastroenteritis accounting for greater than 90% of viral gastroenteritis and approximately 50% of all-cause outbreaks (Atmar 2010). There are 6 norovirus genogroups (GI-GVI) with GI and GII linked to numerous global food and waterborne outbreaks. Of the genotypes associated with these genogroups, a global overview analysis found that approximately 62% of all reported norovirus outbreaks are associated with the GII.4 genotype (Siebenga et al. 2009; Mesquita et al. 2010). However, new strains of norovirus are regularly emerging, evidenced by the recent identification of a GII.17 strain in several locations across the globe (de Graaf et al. 2014; Parra and Green 2015; Han et al. 2015; Mans et al. 2016; Sanchez et al. 2016; Silva et al. 2016).

Norovirus of the Caliciviridae family are small, non-enveloped viruses with a single stranded, positive-sense RNA genome that are environmentally stable as well as resistant to disinfection, heat, and pH changes (Costantini et al. 2006; Le Guyader et al. 2008; Greening et al. 2009; Le Guyader et al. 2010). Norovirus illness is characterized by a mean incubation time of 24 hours, with typical symptoms including the sudden onset of vomiting, diarrhea or both; fever; headache; and muscle aches. Norovirus illness is self-limiting and symptoms usually resolve in 24 and 48 hours (DePaola et al. 2010; Le Guyader et al. 2010). However, it has been observed that individuals with asymptomatic or symptomatic infections can shed norovirus from 8 to 60 days in their feces (Teunis et al. 2015).

With numerous norovirus outbreaks, Quantitative Microbial Risk Assessments (QMRAs) for noroviruses in water and food have emerged (Van Abel et al. 2016). QMRA gathers the best available information to understand the potential risks from microbial exposures and includes 4 steps of (1) hazard identification, (2) exposure assessment, (3) dose-response assessment, and (4) risk characterization (Hunter et al. 2003; Haas et al. 2014).

Exposure Assessment

Some characteristics of norovirus must be considered when developing a QMRA including innate susceptibility and acquired immunity of the host population, the diversity and infectivity of the virus, and the potential for aggregation.

Concentration of Norovirus

Norovirus remains recalcitrant to culture by routine tissue culture methods. As a result, detection and quantification of viral titer in clinical and environmental samples is typically determined by reverse transcription polymerase chain reaction (RT-PCR). qPCR-based methods are important tools because they facillitate quantitative estimates of the concentration of pathogen in water in genomic equivalent copies, RT-PCR units, or PDU (PCR-detectable units). However, this molecular-based approach only measures the physical presence of the pathogens and cannot differentiate as to whether the detected signal represents infectious or inactivated viruses, which creates great uncertainty regarding the human health risk significance (Topping et al. 2009; Girones et al. 2010; Bambic et al. 2011; WHO 2011). Ideally, a quantitative comparison between qRT-PCR and cell culture for norovirus enumeration would be performed to estimate the relationship between genome copies and cell culture infectious units, but until that can routinely be done, the estimated dose and risk estimates must be carefully considered (Teunis et al. 2008; Schijven et al. 2011; Atmar et al. 2014). Additionally, factors such as the method of collection, varation between RT-PCR methods, and inhibitory matrix effects can confound accurate quantification.

Lack of Cell Culture, not 100% Infectivity

Without knowing the fraction of virons that are infectious, the estimated dose is difficult to interpret. For norovirus, the dose may be inflated by the presence of non-infectious viral genomes resulting in underestimatation of risk due to the infectious virions. This could result in observed infections occurring at lower doses when only a few infectious particle may be present (Atmar et al. 2014). Without cell culture infectivity methods, assumptions must be made about the fraction of infectious viruses detected (WHO 2011). In the human challenge studies performed to date, the ratio of total to infectious virus particles in the inoculum used for the dose-response parameterization was unknown. Further in most circumstances, the ratio of infectious viruses is not discerned in exposure matrices, nor is it likely to remain constant do to degradative mechanisms. Thus when performing a QMRA for norovirus, an assumption must be made to decide if the fraction of total to infectious genomes is the same as in the human challenge study (Schoen and Ashbolt 2010).

Typical assumptions are that the majority of the norovirus inocula are infectious. It has been noted that the apparent high infectivity of norovirus in PCR units suggests that there cannot be a large fraction of uninfectious or defective genome copies and it was concluded that potentially at least half of the virus particles (virions) in both 8fIIa and 8fIIb were infectious to humans (Teunis et al. 2008; Thebault et al. 2013). Authors of an oyster outbreak analysis assumed the fraction of viable virus was identical between all outbreaks to eliminate issues of the degree of infectivity. The authors also stated this issue must be addressed in future QMRAs (Thebault et al. 2013). Importantly, the ratio of infectious viruses is likely to vary dependent on the particular exposure scenario (e.g., the ratio in a foodborne outbreak may be very different from a waterborne outbreak due to differences in the level of treatment or disinfection experienced by the viruses).

Several of the issues regarding quantification of noroviruses both in clinical samples and from environmental samples may soon be addressed by recently described tissue culture methods. Historically, determination of norovirus infectivity has been attempted in numerous tissue lines with limited to no success (Ha et al. 2016). However, two novel approaches have recently been described, one utilizing B cells and the other utilizing stem cell-derived human enteroids (Jones et al. 2015; Ettayebi et al. 2016). Both of these methods are critical breakthroughs and will facillitate availability of viral stocks and the study of norovirus replication and mechanisms of infectivity. However, the methods are still complicated and currently show only modest viral output, thus limiting there current application for determination of viral infectivity.

Secretor Status, Innate Susceptibility, and Acquired Immunity

Innate susceptibility, which is related to secretor status, is not fully understood. In general, susceptibility to norovirus infection is dependent on genetic characteristics. Initiation of norovirus infection requires the attachment of the virus to the host receptor, which is dependent on binding of the virus to histo-blood group antigens (HBGAs). HBGAs are complex carbohydrates present on mucosal epithelial, genitourinary, and digestive tracts that are coded or synthesized from 3 families (1) ABO (A/B enzymes), (2) Lewis (FUT3), and (3) secretor (FUT2). Individuals who can secrete the FUT2 enzyme are designated secretor positive (Se+) and have increased susceptibility to norovirus infections (Marionneau et al. 2002; Lindesmith et al. 2003; Shirato 2011; Tan and Jiang 2014; Currier et al. 2015; Lopman et al. 2015; Payne et al. 2015). Individuals who have a non-sense mutation cannot secrete the antigen and are designated secretor negative (Se-). In general, norovirus binding to intestinal epithelial cells occurs in individuals with secretor phenotype (Se+), who are susceptible to norovirus infection. Binding is not observed to cells from non-secretors (Se-), so those individuals are classified as not susceptible (Tan and Jiang 2014; Lopman et al. 2015; Payne et al. 2015); however, the ability of norovirus to infect non-secretors is somewhat genotype-dependent.

Non-secretors demonstrate resistance to specific norovirus genotypes in challenge studies including GI.1 and GII.4 (Lindesmith et al. 2003; Frenck et al. 2012). In fact, in GI.1 human challenge studies no Se- individuals were infected no matter how large the dose (Lindesmith et al. 2003; Teunis et al. 2008; Atmar et al. 2014). However, it should be noted that Se- individuals have been infected by GI.8 (Shirato et al. 2008) and at least one Se- individual became ill during a GII.4 human feeding trial (Frenck et al. 2012). This raises the question as to whether the secretor status for these individuals could have been misclassified or whether secretor status is only part of the picture. Still, epidemiology studies have demonstrated conflicting results about the protectiveness of Se- status. A few studies demonstrated that Se- individuals have some susceptibility to genotypes GI.1, GII.3, and GII.4 (Carlsson et al. 2009; Van Trang et al.

2014; Currier et al. 2015). In contrast, an outbreak study in China found that Se- individuals did not develop infections from either GII.4 or GII.3 (Tan et al. 2008). Yet another study in Sweden observed that non-secretors were equally likely to be infected by GI.3 as secretors (Nordgren et al. 2010). Moreover, GI.1 and GII.4 infections were detected in Se+ children while non-GII.4 infections were detected in Se- children (Lopman et al. 2015).

There is also evidence that norovirus susceptibility may be dependent on blood group as well as the strain of norovirus circulating and it has been shown that the binding of norovirus varies by ABO HBGAs where the specific norovirus genogroups infect only a subset of the population depending on the ABO HBGA expression (Harrington et al. 2002; Lindesmith et al. 2003; Costantini et al. 2006; Le Guyader et al. 2010). Norovirus GI binding to blood group A and O is high, but binding to blood group B was observed to be poor and evidence has demonstrated that individuals expressing O-type HBGAs are more susceptible to GI.1 genotype infections than A and B-types (Hutson et al. 2002; Marionneau et al. 2005; Debbink 2015). Specifically, for GI.1, it was observed that Se+ individuals were of blood group O and A while Se- individuals were of blood group B (Hutson et al. 2002; Lindesmith et al. 2003; Shirato 2011; Atamr et al. 2014). In contrast, GII.4 interacts with all ABO HBGAs (i.e., binding was observed) and infection occurred in all A, B, and O blood groups including 5 individuals with blood group B (Frenck et al. 2012). Also, a norovirus GII oyster outbreak analysis found that individuals in blood group B and O demonstrated illness while individuals of blood group A were less likely to have diarrhea suggesting that binding to blood group A was hindered (Le Guyader et al. 2010). A GII.4 human challenge study found that people of all ABO blood groups had increased susceptibility to GII.4 infection and it was surmised that the increased susceptibility of any ABO blood group to GII.4 could explain why this genotype is the most commonly circuiting genotype worldwide (Hutson et al. 2002; Frenck et al. 2012). Thus, in outbreaks where multiple strains of norovirus are present, the impact of ABO HBGAs on susceptibility will vary, i.e., a specific fraction of the population may be susceptible to a particular norovirus strain, but the fraction will likely be different depending on the norovirus strains (Lindesmith et al. 2003; Teunis et al. 2008; Le Guyader et al. 2010). Ultimately, many HBGA-binding patterns for norovirus have not been defined yet because of the high diversity of norovirus genotypes as well as a lack of data and inconsistent binding results (Tan and Jiang 2014).

Additionally, new evidence has emerged that innate susceptibility is ancestry dependent. This suggests that what is known about norovirus susceptibility in the caucasian population may not fully translate to other ethnic groups (Debbink 2015). In studies of the caucasian populations of the United States and Europe, approximately 70-80% of the population carry FUT2 and are known as secretors (Se+) (Kelly et al. 1995; Currier et al. 2015). In the Asian and African populations, similar percentages of the population carry FUT2 and would thus be secretor positive (Ferrer-Admetlla et al. 2009; Debbink 2015). The secretor status of the Meso-American/Hispanic population is not well-characterized with studies indicating that 2% of the population do not carry FUT2 (Se-) (Payne et al. 2015), 12% of the population in Ecuador were non-secretors (Se-) (Lopman et al. 2015), and 4% of the of Hispanic children in the US were non-secretors (Se-) (Currier et al. 2015).

Finally, acquired immunity is another consideration for norovirus dose-response that is not completely understood. Essentially, short-term, genotype-specific immunity has been observed where acquired immunity from one genogroup or genotype can confer a modest level of protection. This acquired immunity is not permanently sustained lasting from 6 months to more than 4 years, so reinfection can occur throughout life (Saito et al. 2013; Sakon et al. 2014; Payne et al. 2015). In contrast, a recent study of a birth cohort of children in Ecuador did not observe genogroup-specific immunity in that GI and GII infections did not decrease following a previous infection by the same genogroup (Lopman et al. 2015).

Dose-Response Assessment

Dose-response assessment is a key component of QMRA because it links the exposure to a hazardous agent with the health outcome. The choice of dose-response model is important for QMRA (Teunis and Havelaar 2000). A dose-response model takes the form of a mathematical function and an infinite number of these functions exist. Ideally, selection of an appropriate mathematical function to describe dose-response involves functions that are biologically plausible and demonstrate goodness of fit (Haas et al. 2014). For norovirus, multiple models have been described for use in QMRA including the $_1F_1$ and $_2F_1$ hypergeometric and fractional Poisson models (Teunis et al. 2008; Bambic et al. 2011; Messner et al. 2014; Schmidt 2015; Van Abel et al. 2016).

Current Published Human Challenge and Outbreak Data Used in Dose-Response Model Development

All available norovirus human challenge data fed to Se+ individuals are summarized in Table 1. Dose-response data collected from outbreaks are presented in Table 2.

Norovirus Human Challenge Data: Multiple Dose Levels

Two norovirus human challenge studies with multiple dose levels have been reported in the literature by Teunis and Atmar. In the Teunis study, two inocula, 8fIIa and 8fIIb (both genotype GI.1), were fed to human volunteers and the outcomes of infection and illness were recorded (Teunis et al. 2008). The 8fIIa inoculum prepared from the original Norwalk virus isolate in 1971 was used in both a moderate to high dose challenge experiment as well as a low dose challenge experiment. The 8fIIb inoculum was prepared in 1997 from the stool sample of an infected subject from the first challenge experiment and was used in a 3rd challenge experiment. The 8fIIa inoculum was fed to 75 individuals at doses ranging from 3.24 to $3.24*10^8$ genomic equivalent copies (gec) and the 8fIIb inoculum was fed to 33 individuals at doses ranging from $6.92*10^5$ to $2.08*10^7$ gec.

Table 1. Norovirus Dose-Response Data for Se+ Volunteers (adapted from Messner et al. 2014)

group	Genotype (inoculum)	Dose (RT-PCR)	Dose (gec)	Infected (#)	Total Se+ (#)	P_{inf}	Inocula Media	Reference
Multiple Dose Levels								
GI	1 (8fIIa)		$3.24*10^0$	0	8	0	Not described	Teunis et al 2008[a]
GI	1 (8fIIa)		$3.24*10^1$	0	9	0		
GI	1 (8fIIa)		$3.24*10^2$	3	9	0.33		
GI	1 (8fIIa)		$3.24*10^3$	2	3	0.67		
GI	1 (8fIIa)		$3.24*10^5$	7	8	0.88		
GI	1 (8fIIa)		$3.24*10^6$	3	7	0.43		
GI	1 (8fIIa)		$3.24*10^7$	2	3	0.67		
GI	1 (8fIIa)		$3.24*10^8$	5	6	0.83		
GI	1 (8fIIb)		$6.92*10^5$	3	8	0.38		
GI	1 (8fIIb)		$6.92*10^6$	14	18	0.78		
GI	1 (8fIIb)		$2.08*10^7$	1	1	1.00		
GI	1	0	0	0	7	0	Sterile water	Atmar et al 2014[b]
GI	1	0.48	192	1	13	0.08	Sterile water	
GI	1	4.8	$1.92*10^3$	7	13	0.54	Sterile water	
GI	1	48	$1.92*10^4$	7	8	0.88	Sterile water	
GI	1	4800	$1.92*10^6$	6	7	0.86	Sterile water	
Single Dose Level								
GI	1 (8fIIb)		$1.0*10^4$	7	15	0.47	Unprocessed oysters	Leon et al 2011
GI	1 (8fIIb)		$6.5*10^7$	10	13	0.77	Groundwater	Seitz et al 2011[c]
GII	4	$5*10^4$	$2*10^7$d	16	23	0.70	Sterile water	Frenck et al 2012
GII	4	4400	$1.76*10^6$d	30	48	0.63	Sterile water	Bernstein et al 2014[e]

[a] adapted from Teunis et al. 2008 and Bambic et al. 2011 to reflect corrected first 4 dose levels, [b]Estimated 1 RT-PCR unit ≈ 400 gec, [c]The Seitz data were pooled (for storage times ranging from 0 to 61 days) and an assumption of no virus inactivation over the storage time, [d]Assumed the same conversion rate as Atmar et al. 2014, (ex. Messner 2014: $5*10^4$ RT-PCR units *400 gec = $2*10^7$ gec), [e]Data shown are for the placebo group

Subsequent screening of participants for secretor status determined that 22 of the 8fIIa and 6 of the 8fIIb subjects were actually Se- and immune to norovirus, so the dose-response assessment was restricted to only the remaining 53 8fIIa and 27 8fIIb Se+ subjects. The first four dose levels for

8fIIa in Table 1 ($3.24*10^1$ to $3.24*10^4$ gec) were updated because they were incorrectly reported as a factor of 10 too high in the original publication (Bambic et al. 2011; Messner et al. 2014). It should be noted that the subsequent dose-response analysis was done with the correct doses (Teunis et al. 2008).

Table 2. Norovirus Dose-Response Data from Outbreak Studies (adapted from Thebault et al. 2013)

Geno-group	type	NoV Contam (gec/oyster)	Ill (#)	Total (#)	Se+/ Se- (#)	Attack Rate	Outbreak Year	Outbreak Reference[a]
GII	4	18-955-37-0[b]	23	34	27 Se+/ 6 Se-[c]	0.68	2008	Le Guyader et al 2010
GII	n/a	1100[d]	11	27	n/a	0.41	2006[a]	Le Guyader et al 2008
GI	n/a	2300[d]						
GI	n/a	275-6783	2	2	n/a	1	2006	n/a[e]
GII	4, 8	25	21	36	n/a	0.58	2002	Le Guyader et al 2006
GII	4, 9	125						
GI	4	25						
GI	1	85-237	4	4	n/a	1	2000	Le Guyader et al 2003

[a]Publication where the outbreak was further described, [b]4 discrete oyster samples from the same outbreak, [c]Secretor phenotype was identified for 33 of 34 ill individuals, [d]Individual values reported for Shellfish sample #140; reported in Table 2 of outbreak reference, [e]Results of this outbreak did not appear in any of the publications, obtained from outbreak database

In a second study by Atmar, dose levels of GI.I challenge inoculum (lot 42399) prepared from the liquid feces of a challenge participant ranging from 0.48 to 4800 RT-PCR units ($1.92*10^2$ to $1.92*10^6$ gec) were fed to study participants and the outcome of infection was recorded (Graham et al. 1994; Atmar et al. 2011; Atmar et al. 2014). Eligible participants were limited to Se+ individuals. One RT-PCR unit of Norwalk virus was estimated to contain approximately 400 gec based on a most probable number (MPN) approach using 5 amplified replicates of 10-fold dilutions of nucleic acid (1/10, 1/100, 1/1000) that were determined to be positive or negative by Southern hybridization (Le Guyader et al. 2003).

The reported 400 gec value is considerably higher than other published values. A previous study using an MPN RT-PCR approach estimated that 1 RT-PCR unit was approximately 10 to 50 gec for norovirus (Le Guyader et al. 2003). Another shellfish study found that approximately 420 gec were present in a 3200-fold dilution of extracted stool sample, which represents 10 RT-PCR units; thus, 1 RT-PCR unit was equal to approximately 42 gec (Atmar et al. 1995). A third study compared endpoint titration to quantitative RT-PCR (RT-qPCR) for 8fIIa and 8fIIb to estimate the RT-PCR detectable unit (PDU) to genomic equivalent copies (gec) ratio. Briefly, endpoint titration involved serially diluting the inocula in replicates to find the highest dilution for which a visible PCR amplicon could be observed and assigning this as 1 PDU. RT-qPCR estimated the gec values from the CT values, which are defined as the PCR cycle where the fluorescent signal can be detected, and known copy numbers using the norovirus RNA standard curve. For 8fIIa, endpoint titration estimated $7.6*10^7$ PDU/ml and RT-qPCR estimated $2.9*10^9$ gec/ml, which results in a ratio of approximately 38. For 8fIIb, endpoint titration estimated $5.9*10^7$ PDU/l as compared to $7.2*10^8$ gec/ml with RT-qPCR, which is a ratio of approximately 12 (Liu et al. 2010). Ultimately, there is undoubtedly considerable natural variability in the adjustment factor between RT-PCR units and gec, a factor that should be accounted for in the uncertainty of norovirus dose-response.

Norovirus Human Challenge Data: Single Dose Level

In addition to dose-response human challenge studies, a few other peer-reviewed articles were identified that provided a dose of norovirus and an outcome of infection. In one study, the effect of high hydrostatic pressure processing of oysters on norovirus inactivation was assessed by feeding subjects processed or unprocessed oysters to determine whether infection occurred (Leon et al. 2011). In another study, the long-term infectivity of norovirus was examined by feeding subjects groundwater that had been spiked with norovirus and stored for certain lengths of time (Seitz et al. 2011). In the Frenck et al. study, authors sought to perform the first human challenge study of Norovirus GII.4 and to evaluate the relationship between infection and HBGAs (Frenck et al. 2012). Another study challenged patients with Norovirus GII.4 to test the efficacy of a newly developed GI.1/GII.4 vaccine (Bernstein et al. 2015).

Outbreak Data

Outbreak data can also provide an important source of information on the infectivity of norovirus GI and GII. One study collected data from oyster-associated outbreaks on the genogroup and type, attack rate, dose in oyster, and secretor status if available (Table 2) (Thebault et al. 2013). In two outbreaks, contamination by both genogroups was observed (Le Guyader et al. 2006; Le Guyader et al. 2008), while the others only observed one genogroup (Le Guyader et al. 2003). Only one outbreak collected information on secretor phenotype (Le Guyader et al. 2010).

Considerations of Norovirus Dose-Response Models

Some characteristics of norovirus are considered in dose-response data including innate susceptibility and acquired immunity, the diversity and infectivity of the virus, and the potential for aggregation.

Innate Susceptibility Due to Secretor Status and Acquired Immunity

Norovirus human challenge studies are also limited to Se+ healthy adults. Thus, the assumption in most QMRAs is that the exposed population are Se+ individuals.

GI and GII Genogroups

While outbreaks can occur from either GI or GII genogroup; the multi-dose level norovirus human challenge study data is predominantly for the genogroup GI.1. Thus the published dose-response models are only for GI. This forces an assumption in risk assessment that both the GI and GII genogroups (and individual strains) are equally infectious (McBride et al. 2013). One study that analyzed norovirus susceptibility from 5 oyster outbreaks found no difference in infectivity between GI and GII strains (ID_{50} values). However, the study did note that large variances were observed for the estimated infectivities such that differences between the genogroups may be observable with more information (Thebault et al. 2013). Also, a vaccine trial mentioned that the genetic diversity between and within genogroups could present a major challenge to vaccine development (Bernstein et al. 2015). To date due to the lack of published information, the relative infectiousness of GI as compared to GII is still unclear (Teunis et al. 2008; Leon et al. 2011; Seitz et al. 2011). Thus, more research needs to be done to understand the potential differences in infectivity of the two commonly circulating strains.

Infectious vs Non-Infectious Virus Particles

In the human challenge studies, the ratio of total to infectious virus particles in the inoculum used for the dose-response parameterization was unknown. Thus, when using norovirus in QMRA an assumption must be made to decide if the fraction of total to infectious genomes is the same as in the human challenge study (Schoen and Ashbolt 2010). Typical assumptions are that the majority of the norovirus inocula are infectious.

The dose-response parameterization only holds true when the ratio of total to infectious virions is consistent and the exposure estimates are based on the same enumeration method as the dose-response enumeration (Teunis et al. 2008; Schoen and Ashbolt 2010; McBride et al. 2013). If the laboratory methods used for pathogen enumeration in the human challenge study differ from those used to enumerate the pathogen in the source water, then an analysis must be done to ensure that the results are harmonized. Harmonization involves modifying environmental media measurements by an adjustment factor. The basis for the adjustment factor may be due to inhibition in the measurement due to exposure matix effects, diffience in expected infectivity ratio, or variability in the methods of detection. In a QMRA from exposure to stormwater-borne pathogens, authors developed harmonization rules for 7 pathogens including norovirus. The authors found that the dose-response and environmental studies both used qPCR for enumeration, but different target sequences were used. This resulted in differences in critical threshold standard curves, so the human challenge (clinical) study data was adjusted by dividing by 18.5 (McBride et al. 2013). This is an example of how harmonization of the methods used to enumerate the exposure pathogen concentration and the dose-response concentration can be performed in QMRA. However, this adjustment factor should not be extrapolated to future QMRAs. Instead harmonization should be carried out and thoroughly defended for each specific QMRA.

Aggregation

In a human challenge study the inoculum (8fIIa) was described as having been stored for many years in a stabilizing solution that was high in protein and thought to promote viral aggregation. Also, the inoculum was observed to be 'sticky" and viral aggregation was observed by electron microscropy. Investigators indicate that they were unable to easily dissociate the inoculum by dilution. As a consequence, the study authors concluded that the dose must be considered as a mixture of aggregates as opposed to individual viruses and be represented as a Poisson-stopped log-series distribution (Teunis et al. 2008). Some agree that aggregation should be considered in norovirus dose-response and subsequent models also accounted for aggregation (McBride 2014; Messner et al. 2014). However, recently, the addition of the aggregation parameter has been called into question being labeled a tuning parameter for the model (Schmidt 2015). It should be noted that no accurate measurement of the aggregate size or amount of viral aggregation has been performed on the 8fIIa inoculum. In contrast, the 8fIIb inoclum was described as being not aggregated because it came from a more recent second human passage (Teunis et al. 2008; Messner et al. 2014). However, no electron micrograph of 8fIIb was provided.

The isoelectric point (IEP), the pH at which a particle has a net surface charge of zero, is an important value that can providc information of the surface charge of the virus and its influence on sorption behavior. When considering viral aggregation, both the IEP of the viral particle and the pH of the suspending medium should be considered because aggregation has been observed to increase as the pH moves farther below the IEP (Langlet et al. 2007; Michen and Graule 2010). For example in a pH aggregation study where the IEP of MS2 bacteriophage was assumed to be 3.9 (others report an average of 3.5 with a standard deviation of 0.6), as the pH decreased below the IEP electrostatic interactions increased and more aggregation was observed (Overby et al. 1966; Langlet et al. 2007).

Another study found that as the pH decreased (from 3.9 to 2.5) aggregate size increased to 2 to 6 μm (2000 to 6000 nm) (Langlet et al. 2007). When the pH was larger than the IEP, aggregation of MS2 bacteriophage was not observed (Langlet et al. 2008). For poliovirus (IEP: 6.5 to 8.3), aggregation was noticeable at a pH of 6 but was so substantial at a pH of 5 or 3 such that only a few individual particles were observed to not be in aggregates (Floyd and Sharp 1977; Michen and Graule 2010). For adenovirus (IEP: 3.5-4.0) increasingly larger aggregates were observed by electron microscopy when the pH was lowered (Johnson and Bodily 1975; Wong et al. 2012). Overall, pH and IEP should be considered when making assumptions about the aggregation parameters in dose-response models.

The IEP for norovirus GI as reported in the literature ranges from 5.9-6.0 (Goodridge et al. 2004; Michen and Graule 2010). In the human challenge studies, Se+ individuals were fed inoculum that contained virus determined to be "sticky" due to the proteins and the pH of the inoculum is unknown. The potential for the ingested dose to remain aggregated depends on the pH of the stomach as well as factors in how the dose was administered. Once the dose is ingested, the aggregation may impact how norovirus behaves relative to the ingestion of the same number of disaggregated virions. Aggregates could facilitate or prevent infectious particles from binding in the gut and leading to infection (Teunis et al. 2008). The pH of the human gut is quite acidic, so the pH would be less than the IEP and strong, stable aggregation could be maintained following ingestion of the inoculum sample. However, human challenge study protocols often involve subjects ingesting sodium bicarbonate prior to and following ingestion of the virus inoculum to neutralize gastric acidity (Leon et al. 2011; Seitz et al. 2011; Frenck et al. 2012; Atmar et al. 2014). Therefore, consideration should be given to how the viral aggregate behaves if the stomach acid is neutralized before ingestion of the dose because potentially there could be an impact on norovirus aggregation. The dose of norovirus would not change but how the viral particles act inside the gut may change because the viral particles may no longer be in aggregates.

The pH of environmental media should also be considered when making assumptions about viral aggregation in QMRA. For example, recreational water or drinking water would not be acidic and would have a neutral pH, suggesting limited viral aggregation. However, it has been observed that small changes in the pH of pore space water may strongly influence the electrostatic character and ultimately aggregation for Norwalk virus (Redman et al. 1997). Currently, assumptions about aggregation are already made in dose-response because viruses are often presumed to be disaggregated in environmental media, such as water (Schoen and Ashbolt 2010; McBride et al. 2013). Because the pH of water is greater than the IEP of norovirus, the virus may be assumed to be disaggregated in environmental media. Another consideration is that norovirus in environmental samples may be in a very different aggregation state than the aggregated inoculum, so incorporating viral aggregation into the dose-response model was important to ultimately consider the infectivity of the virus as a single virus particle (Teunis et al. 2008). However, without measurements or methods to predict norovirus aggregation, it is still uncertain how important the viral aggregation term is in dose-response.

Norovirus Dose-Response Mathematical Models

Traditionally, the exponential and approximate beta-Poisson dominated microbial dose-response because of the ease of implementation, but for norovirus the hypergeometric functions were proposed in order to account for possible virus aggregation (Teunis et al. 2008). The following describes all available dose-response models for norovirus.

Dose-Response Models with Disaggregated Dose

Microbial dose-response is unique because it theoretically takes only a single infectious microorganism to make an individual sick. The single-hit model theory for a disaggregated dose, which dominates microbial dose response, assumes a sequence of events leads to microbial infection: (1)

intake of 1 or more microbes which are distributed as a probability function such as Poisson and (2) successful evasion of at least one microorganism past the host's defensive barriers, which then leads to infection. Individual organisms are assumed to act independently (Teunis and Havelaar 2000; Teunis et al. 2008; Haas et al. 2014). The single hit theory is written as

$$P_{infection} = 1 - (1 - p_m)^n \tag{1}$$

Where n is the integer amount of microorganism ingested and p_m is the probability of a "single hit" or probability of an independently acting viral particle to survive the host's defensive barriers and cause infection (Teunis and Havelaar 2000). When microorganisms are randomly distributed in the inoculum, then the random error in the number contained in a particular volume of inoculum is Poisson distributed. Then the single-hit theory says that probability of infection from the ingestion of a particular mean dose of microorganism is the exponential dose-response relationship

$$P_{infection} = 1 - e^{-p_m * dose} \tag{2}$$

where $p_{infection}$ is probability of infection and characterizes the process (Haas et al. 2014). The exponential model does not take into account the range of human responses to pathogens. Instead, to account for this heterogeneity, p_m is represented by the beta probability distribution function. The beta distribution is a continuous probability distribution with two positive shape parameters (α, β), which control the shape of the distribution (Pearson 1983). The resulting dose-response relationship takes the form of the ${}_1F_1$ hypergeometric function (confluent hypergeometric function), also called the full or exact beta-Poisson and is expressed as

$$P_{infection} = 1 - {}_1F_1(\alpha, \alpha + \beta, -dose) \tag{3}$$

where α and β characterize the variation in the host susceptibility (Abramowitz and Stegun 1965; Teunis and Havelaar 2000; Haas et al. 2014; Schmidt 2015).

An approximation of the $_1F_1$ hypergeometric function when β >>1 and α << β yields a simplification of equation 3 to equation 4 (Furumoto and Mickey 1967). This is called the approximate beta-Poisson equation and is written in the form

$$P_{infection} \approx 1 - \left(1 + \frac{dose}{\beta}\right)^{-\alpha} = 1 - \left(1 + \frac{dose}{N_{50}}\left(2^{\frac{1}{\alpha}} - 1\right)\right)^{-\alpha} \quad (4)$$

where α and β are the dose-response parameters and N_{50} is the median infectious dose or the dose at which 50% of people are infected. The slope parameter, α, reflects an increase in the degree of heterogeneity in the pathogen-host interaction that leads to infection while β is a scale parameter that leads to a shift in the along the dose axis (Haas et al. 1993; Teunis and Havelaar 2000). N_{50} and β are directly proportional and are related by the following equation.

$$N_{50} = \beta * \left(2^{1/\alpha} - 1\right) \quad (5)$$

Dose-Response Models with Aggregated Dose

When viruses are not suspended as individual particles but as aggregates of variable size as was assumed by some dose-response modelers, then the previous assumption that the inoculum was Poisson distributed does not hold. For instance, if viral particles formed large aggregates then dilution of the inoculum which has a particular number of viral particles may actually only contain a few viral aggregates. As a result, if the aggregates are ingested according to a Poisson distribution, then a subject receiving a large dose (>1000 gec) would have a high chance of not ingesting any virus because only a few viral aggregates are present. As virus aggregates may not contain all the same number of virions, the

Poisson distribution does not adequately describe the intake of infective particles and instead the aggregated viral suspension is assumed to be a Poisson stopped logarithmic series distribution where the aggregates are assumed to be Poisson distributed and the aggregate sizes are assumed to be logseries distributed with parameter 'a' (Teunis et al. 2008). Thus, the single hit model for infection from an aggregated dose becomes

$$P_{infection} = 1 - \left(1 + p_m \frac{a}{1-a}\right)^{dose/\log(1-a)} \quad (6)$$

where p_m is the single hit infectivity and 'a' is the parameter of the logarithmic series aggregate size distribution (Teunis et al. 2008). This model is the aggregated analog of the exponential model.

Aggregation is accounted for in dose-response models through an aggregation parameter ('a' or μ_a). The 'a' parameter characterizes the aggregate size distribution ($0 \leq a < 1$) with an 'a' of zero indicating completely disaggregated virus and higher values of 'a' corresponding to higher degrees of aggregation. The 'a' parameter is determined from μ_a, which describes the average aggregate size, and is defined as the following (Messner et al. 2014).

$$\mu_a = \frac{-a}{(1-a)\ln(1-a)} \quad (7)$$

Values of the parameter μ_a have been proposed based on model fitting, but it has subsequently been shown that such approaches may not yield statistically valid estimates of the aggregation parameter (Schmidt 2015).

When host-to-host variability is expressed as the beta distribution with parameters α and β, then the dose-response model includes a ${}_2F_1$ hypergeometric function (Gaussian hypergeometric function) (Koepf 1998) and the microbial dose-response model is expressed as

$$P_{infection} = 1 - {}_2F_1(\alpha, \frac{dose*(1-a)}{a}, \alpha + \beta, \frac{-a}{1-a}) \quad (8)$$

where α and β are the estimated dose-response parameters and 'a' is the viral aggregation parameter (Abramowitz and Stegun 1965; Teunis et al. 2008).

The $_2F_1$ hypergeometric function only holds true when |-a/(1-a)| < 1, otherwise the Pfaff transformation should be applied and is expressed as the following equation (Koepf 1998; Weisstein 2014).

$$P_{infection} = 1 - \left[\left({_2F_1}\left(\beta, \frac{dose(1-a)}{a}; \alpha+\beta; a\right)\right)\left(\frac{1}{1-a}\right)^{\left(\frac{-dose(1-a)}{a}\right)}\right] \quad (9)$$

At very large doses (>33,323 viruses ingested/person/day when a = 0.9997), it has been observed that the Pfaff transformation cannot come to a solution ("fails") in R software (R Foundation for Statistical Computing, Vienna, Austria) when using the package 'gsl' (Hankin 2013). Therefore, at doses >33,323 gec/person/day it was assumed that the $_1F_1$ hypergeometric dose-response model provides an adequate approximation of the $_2F_1$ hypergeometric (Barker et al. 2013a; Barker et al. 2013b; Mok et al. 2014). In Mathematica (Wolfram Research, Campaign, IL, USA), the $_2F_1$ hypergeometric did not fail at large doses. Instead, the results for the $_2F_1$ with and without the Pfaff transformation were identical to the results obtained from the R software using the $_2F_1$ and the $_1F_1$ approximation (results not shown).

A simplification of the $_2F_1$ hypergeometric function by eliminating the two-parameter beta distribution instead employing a Bernoulli variable that characterizes a host as perfectly susceptible or perfectly immune yields another proposed model for aggregated dose called the fractional Poisson. This assumption was made because when alpha and beta are much less than 1, then the beta-distribution's probability density is massed near either extreme of 0 or 1. In this case, the probability of infection, p_m, is a Bernoulli random variable with either a value of 1, when individuals are perfectly susceptible with probability, P, or a value of 0, when individuals are perfectly immune with probability, 1-P. As a Bernoulli variable, p_m will always be 0 or 1 and nothing in between. The fractional Poisson still contains a parameter to account for viral aggregation and can be expressed as

$$P_{infection} = P * (1 - e^{-dose/\mu_a}) \tag{10}$$

where P is the fraction of subjects who are fully susceptible and μ_a is the mean aggregate size (number of viruses per aggregate) (Messner et al. 2014).

A new dose-response relationship has been proposed by Schmidt (2015) that generalizes the $_2F_1$ hypergeometric model to allow for immunity among secretors or generalizes the fractional Poisson to not assume full susceptibility among the non-immune because there may also be unexplained apparent immunity in the Se+ population that should be addressed. This new model called the aggregated exact beta-Poisson with immunity addresses the issues of immunity and susceptibility and is expressed as

$$P_{infection} = (1 - \varphi) * \left[\left(1 - \theta \cdot {}_2F_1\left(\beta, \frac{dose(1-a)}{a}; \alpha + \beta; a\right)\right)\right] \tag{11}$$

$$\theta = e^{\frac{dose(1-a)*\ln(1-a)}{a}} \tag{12}$$

where φ is the immunity parameter, α and β are the estimated dose-response parameters, and ‘a’ is the viral aggregation parameter. This parameterization of the $_2F_1$ hypergeometric is different from equation (9) for ease of numerical summation (Schmidt 2015).

Figure 1 summarizes all the available dose-response models and approximations published for norovirus. The available models are divided as to whether a disaggregated dose (dark shade) or aggregated doses (light shade) is assumed.

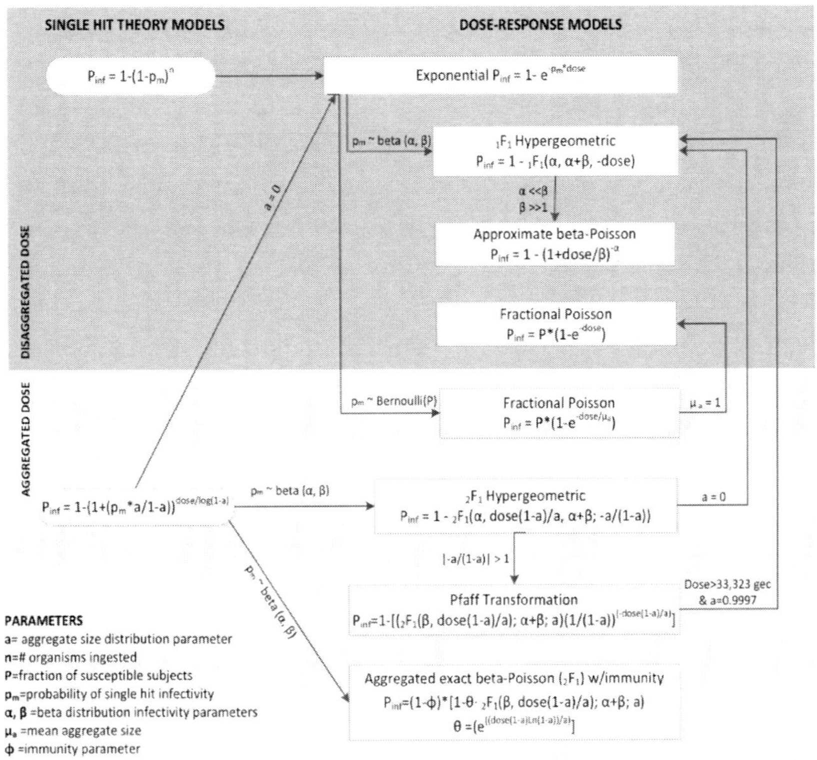

Figure 1. Published dose-response models, approximations, and assumptions for Norovirus.

Parameterization of the Norovirus Dose-Response Models

The model parameters and ID_{50} values for multiple dose-response models are summarized in Table 3. All models and ID_{50} values were determined from human-challenge data, so all values are for Se+ individuals. The numerous best-fit dose-response parameters were collected from published literature where the fitting method used was described in detail (Teunis et al. 2008; Atmar et al. 2014; McBride 2014; Messner et al. 2014; Schmidt 2015) except for the approximate beta-Poisson, which was estimated directly by code from CAMRA (Center for Advancing Microbial Risk Assessment, Michigan State University) previously detailed (Bartrand et al. 2008) and written in R software (R Foundation for Statistical Computing, Vienna, Austria). The following data were used to fit the listed dose-response models (data available in Tables 1 and 2): 8fIIa inoculum alone (Teunis et al. 2008), the 8fIIa+b inocula combined (Teunis et al. 2008), Atmar lot 42399 alone (Atmar et al. 2014), GI outbreak data (Thebault et al. 2013), and a pooled (GI, GII) human challenge dataset that included multiple published studies (Teunis et al. 2008; Atmar et al. 2011; Seitz et al. 2011; Frenck et al. 2012; Messner et al. 2014). Some dose-response datasets combined all genotype GI.1 inocula data points of 8fIIa (aggregated) and 8fIIb (disaggregated), which accounted for uncertainty and variability in aggregation. Other dose-response datasets combined GI and GII, which incorporated the variability between norovirus strains in the average fit estimate. The best fit dose-response parameters are identified by the method of fitting in the references (CAMRA) otherwise the reference where the parameters were originally published is included. Each dose-response model has an associated abbreviation that briefly identifies the dose-response model, dose-response dataset (T for Teunis, A for Atmar, Th for Thebault, or P for Pooled), genogroup, and genotype. For example, the $_1F_1$ hypergeometric model fit to the Atmar dose-response dataset would be abbreviated 1F1_AGI. For dose-response models fit to aggregated data but used with a disaggregated assumption then an x was added to the abbreviation (i.e., 1F1x_AGI). For the model that incorporates immunity, an 'i' was added to the abbreviation. The best-fit dose-response parameters for Atmar (2014) data presented are for adjusted and unadjusted data. Adjusted data converted the RT-PCR units to gec using an adjustment factor of 1 RT-PCR unit equals 400 gec.

The majority of the published dose-response models (Table 3) for both disaggregated (a) and aggregated dose (b) are shown in Figure 2 along with all published multi- and single-dose level human challenge data (Table 1). Two single dose level data points for GII.4 (Frenck et al. 2012; Bernstein et al. 2015) were adjusted to gec because RT-PCR units were reported. Additionally, the 1F1_AGI model was subsequently adjusted from RT-PCR units to gec because the model was fit to RT-PCR data.

Table 3. Summary of Estimated Parameters and ID_{50} Values for Available Norovirus Dose-Response Models for Se+ Individuals

Abbrev	DR Model	DR data Ref	Geno- group	Geno- type	DR Model Parameters α	β	P	'a'	μ_a	ID_{50} (gec)	DR fit Ref
Disaggregated Dose											
LR_AGI(bgO)[a]	Log Regression	Atmar et al 2014	GI	1						1,320 (440-3,760)[c]	Atmar et al 2014
LR_AGI(bgA)[b]	Log Regression	Atmar et al 2014	GI	1						2,800 (290-25000)[c]	
BP_AGI[d]	Approx BP	Atmar et al 2014	GI	1	0.349	357.1				2,249	CAMRA[e]
BP_TGIa	Approx BP	Teunis et al 2008	GI	1 (8fIIa)	0.111	32.79				16,963	
BP_TGIa+b	Approx BP	Teunis et al 2008	GI	1 (8fIIa+b)	0.104	32.29				25,430	
1F1_AGI[f]	1F1	Atmar et al 2014	GI	1	0.46	1.2				1,624[g]	McBride 2014
1F1x_AGI[h]	1F1	Atmar et al 2014	GI	1	0.064	0.018				403	
1F1x_TGIa+b[h]	1F1	Teunis et al 2008	GI	1 (8fIIa+b)	0.04	0.055				18[i]	Teunis et al 2008
1F1_ThGI	1F1	Thebault et al 2013	GI	1 ,4	0.05	0.063				7	McBride 2014
FPx_TGIa+b[h]	FP	Teunis et al 2008	GI	1 (8fIIa+b)			0.686 (0.55-0.79)			~1[j]	Messner et al 2014
FPx_PGIa+bGII[h]	FP	Pooled[k]	GI, GII	1 (8fIIa+b), 4			0.72 (0.63-0.80)			~1[j]	
2F1i_ PGIa+bGII[l]	2F1	Pooled[k]	GI, GII	1 (8fIIa+b), 4	2.910	2734	0.7246[m]	0	1[n]	1,355	Schmidt 2015

Table 3. (Continued)

Abbrev	DR Model	DR data Ref	Geno- group	Geno- type	DR Model Parameters α	β	P	'a'	μ_a	ID_{50} (gec)	DR fit Ref
Aggregated Dose											
2F1_TGIa	2F1	Teunis et al 2008	GI	1 (8fIIa)	5.35E-03	2.51E-03		0.9998	581.9	758	Teunis et al 2008
2F1_AGI	2F1	Atmar et al 2014	GI	1	0.064	0.018		0.9[p]	3.91[q]	1,527	McBride 2014
2F1_TGIa+b	2F1	Teunis et al 2008	GI	1 (8fIIa+b)	0.04	0.055		0.9997	396.4	1,015[i]	Teunis et al 2008
2F1_PGIa+bGII[q]	2F1	Pooledk	GI, GII	1 (8fIIa+b), 4	0.0044	0.0020[f]		0.9999[q]	1,024	1,367	Messner et al 2014
FP_TGIa+b	FP	Teunis et al 2008	GI	1 (8fIIa+b)			0.686[s] (0.55-0.79)	0.9998[q]	579[s] (140-2518)	756	
FP_PGIa+bGII	FP	Pooled[k]	GI, GII	1 (8fIIa+b), 4			0.72[s] (0.63-0.80)	0.9999[q]	1,106[s] (399-2428)	1,305	

[a]Estimated for blood group O,
[b]Estimated for blood group A,
[c]Median and 95% confidence intervals,
[d]DR model fit to adjusted data,
[e]Fit for this paper with CAMRA code
[f]DR model fit to unadjusted data (RT-PCR units),
[g]ID50 value adjusted to gec after fitting,
[h]1F1 parameters are the same as 2F1 or FP DR model parameters (aggregated) but with disaggregated dose assumption (a=0, a=0.0001),

[i]Reported ID50 values for 1F1 of 18-26 gec and for 2F1 1015-1058 gec,

[j]Estimated ID50 with disaggregated assumption was between 1 and 2 gec,

[k]Pooled DR dataset is Teunis 2008, Seitz 2011, Frenck 2012, Atmar 2013,

[l]This model incorporated immunity

[m]Where $P=1-\varphi$ and $\varphi=0.2754$

[n]The aggregated 2F1 model was fit and the best-fit parameter for μa was 1, which suggests disaggregation

[p]Value arbitrarily selected by authors,

[q]Value calculated by equation 7,

[r]Revised paper published incorrect β value, reported table parameter is corrected per communication with Michael Messner, PhD (October 30, 2014),

[s]MLE value (95% credible interval

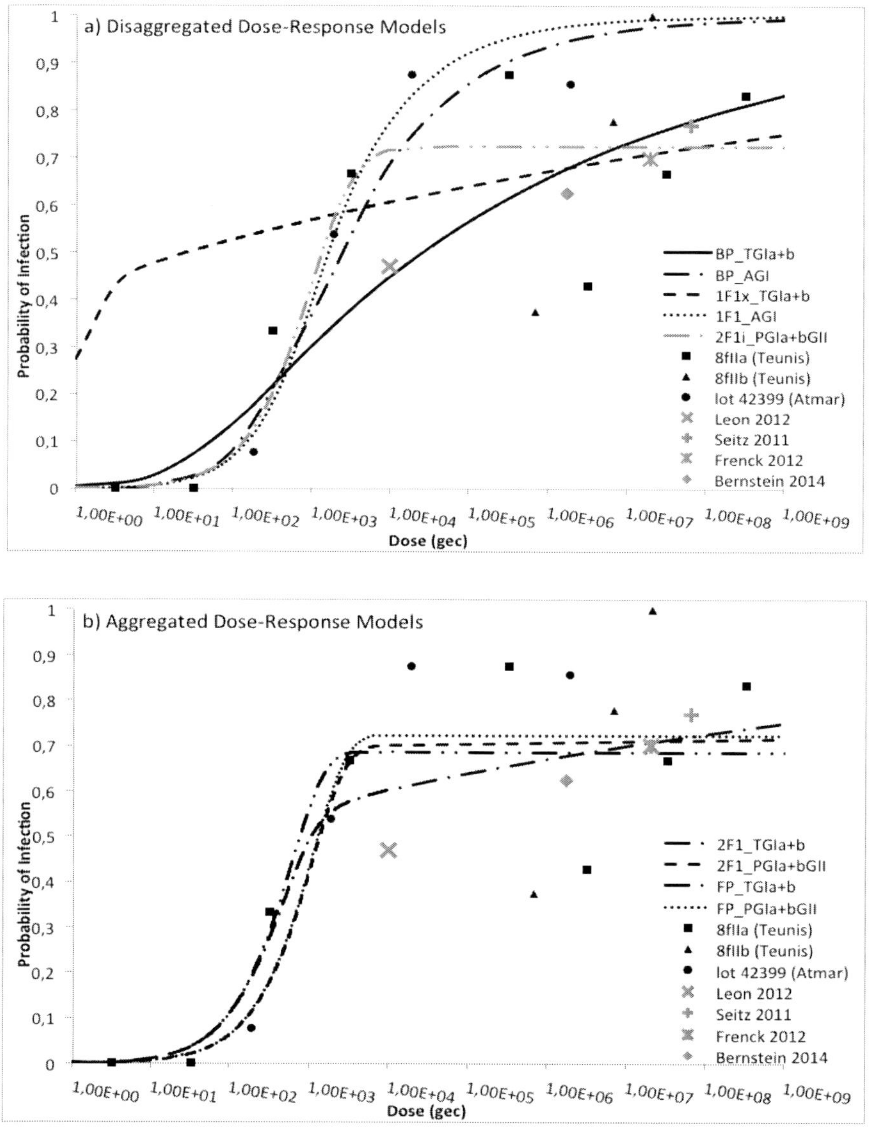

Figure 2. Multiple Dose-Response Models with Published Human Challenge Data (a) Disaggregated Dose Models (b) Aggregated Dose Models (reprinted from van Abel et al., 2016).

Discussion of Difference in Published Dose-Response Models

Probability of Infection Variation across Doses

The disaggregated models demonstrate great disparity in estimated probability of infection values at varying doses (Figure 2a). At low doses (<~1000 gec), BP_TGIa+b, BP_AGI, 1F1_AGI, and 2F1i_PGIa+bGII predict similar probability of infection values (<0.3). In contrast, 1F1x_TGIa+b estimates much higher probability of infection values ranging from approximately 0.3 to 0.5. Only 4 dose levels of data are available at these lower doses including 2 dose levels <100 gec that found no one being infected from norovirus. None of the published dose levels are near the 1F1x_TGIa+b dose-response model.

At doses greater than ~1000 gec, the dose-response models fit to Atmar data, BP_AGI and 1F1_AGI, predict much higher risk estimates than the models fit to Teunis data, BP_TGIa+b and 1F1x_TGIa+b, or to the pooled dataset, 2F1i_PGIa+bGII. Also, none of the disaggregated models predict probability of infection values very close to the disaggregated 8fIIb values and the 1F1x_TGIa+b model is very far from the 8fIIb data points. In contrast, the aggregated dose-response models (Figure 2b) were clustered across all doses, predict very similar risk estimates no matter the dose level, and all leveled off around 0.70 probability of infection at a dose of approximately 5000 gec. The addition of an aggregation parameter allowed the model to remain near to a probability of zero at small doses (<1000 gec) before increasing sharply (Figure 2b). The leveling off of the probability of infection well below 1 suggests immunity within the Se+ population, which is addressed in the 2F1i_PGIa+bGII.

In addition to all the aggregated dose-response models, two disaggregated dose-response models (1F1x_TGIa+b and 2F1i_PGIa+bGII) demonstrate plateauing at around 70% probability of infection. Both of these were initially aggregated models with the 2F1i_PGIa+bGII finding a best-fit value for ‘a’ to be 0 and the 1F1x_TGIa+b forcing the ‘a’ to zero. This assumption made in 1F1x_TGIa+b has not been tested and would only hold true if the α and β are biologically meaningful to describe the host-virus interaction for both individual virions and aggregates. Also, it should be noted that while it is reasonable to use the same beta distribution parameters for either disaggregated or aggregated dose, the values of α and β are dependent on 'a' when fitting the models. This dependency is a result of all parameters being fit simultaneously.

Differences in ID_{50} Values

Differences in estimated ID_{50} values, which is the infectious dose where 50% of the population is infected, were observed. ID_{50} values were determined from human challenge (Table 1) or from outbreak data (Table 2). Human challenge data is often limited to one strain, so the ID_{50} values may not be generalizable to other genotypes. In the case of norovirus, the majority of the human challenge studies use genotype GI.1; therefore, outbreak data is an important source of complementary and contemporary information (Teunis et al. 2008; Thebault et al. 2013). Table 3 summarizes the ID_{50} estimates, which vary from ~1 to 25,000 gec. The majority of the ID_{50} values ranged from approximately 400 to 3000 gec, while some more extreme values were estimated at <20 and >15,000 gec. In fact, over half of the dose-response model scenarios estimated ID_{50} values in the range of 750 to 1600 gec. Differences in the ID_{50} values were observed depending on whether the multi-dose level human challenge data set was used to fit the models, between genogroups and secretor status when outbreak data was available, and whether aggregation was considered.

The different human challenge dose-response datasets yielded different ID_{50} estimates. With the Atmar dataset, an ID_{50} value of 2249 gec was estimated from the approximate beta-Poisson, 1624 gec from the ${}_1F_1$ hypergeometric, and 1527 gec from the ${}_2F_1$ hypergeometric. In addition, this human challenge multi-dose level study used logistic regression to estimate an ID_{50} value of approximately 1320 gec for blood type O and 2800 gec for blood type A (Atmar et al. 2014). In the Teunis multi-dose level human challenge study (Table 1), the ID_{50} estimates ranged from 18 for the ${}_1F_1$ hypergeometric to 758 gec for the fractional Poisson to 1015 gec for the ${}_2F_1$ hypergeometric. The ID_{50} for the ${}_1F_1$ hypergeometric was not obtained directly, but by zeroing the aggregation parameter in the fitted

${}_2F_1$ and the data do not support such a low ID_{50} value. In fact, as observed by others, no one was infected who was administered a dose of 32.4 gec or less and only 3 out of 9 individuals were infected at a dose of 324 gec (Teunis et al. 2008; Atmar et al. 2014). Thus, the low dose ID_{50} values are extrapolated out of the range of data available. Ultimately, more data is needed at low doses to understand what is happening.

Differences in ID_{50} estimates were observed in an investigation of the difference in norovirus infectivity based on genogroup and secretor status. Data from 5 oyster outbreaks (see Table 2) were analyzed in a Bayesian framework to determine the median ID_{50} and 95% credible intervals. The mean dose was assumed to be Poisson distributed (the virus was fully dispersed and homogenously mixed), so the outbreak data were fit to the ${}_1F_1$ hypergeometric function (equation 3). The results for the ID_{50} values are presented in Table 4.

The ID_{50} values varied by secretor status with Se+ individuals having a median ID_{50} value of 7 gec for GI and 1.6 gec for GII. The 95% credible intervals were quite wide ranging from less than 1 to greater than 10^6 gec for both genogroups demonstrating the large variability and uncertainty in norovirus infectivity. For Se- individuals for both genogroups the estimates of the median ID_{50} values were $>10^6$ gec, which supports the idea that Se- are protected from norovirus infection. The authors indicated that no difference in infectivity was detected between genogroups; however, no statistical analysis was performed to determine if the difference was significant. The results showed that Se+ individuals demonstrate a high probability of infection and disease at low doses while Se- individuals possessed a decreased susceptibility to norovirus infection (Thebault et al. 2013).

Whether the dose in the dose-response models was disaggregated or aggregated also yielded different ID_{50} estimates. The ID_{50} estimates for disaggregated models were disparate ranging from approximately 1 to 25,000 gec while for the aggregated models the estimates were more similar ranging from 756 to 1527 gec. The most extreme ID_{50} values were obtained with the disaggregated dose-response models with the highest ID_{50} values being estimated using the approximate beta-Poisson relationship and the lowest by the $_1F_1$ hypergeometric and the fractional Poisson with disaggregated dose. In contrast, the aggregated dose-response models had clustered ID_{50} estimates in a narrow range. Finally, aggregated and disaggregated models with the same dose-response parameters yielded different ID_{50} values.

Table 4. ID_{50} Values for the $_1F_1$ Hypergeometric Obtained from Outbreak Data (adapted from Thebault et al. 2013)

Secretor Status	Geno-		ID_{50} (gec)	
	Group	Type	Median	(95% CI)
Se+	GI	1, 4	7.1	(0.73 to >10^6)
Se+	GII	4, 8, 9	1.6	(0.74 to >10^6)
Se-	GI	1, 4	>10^6	
Se-	GII	4, 8, 9	>10^6	

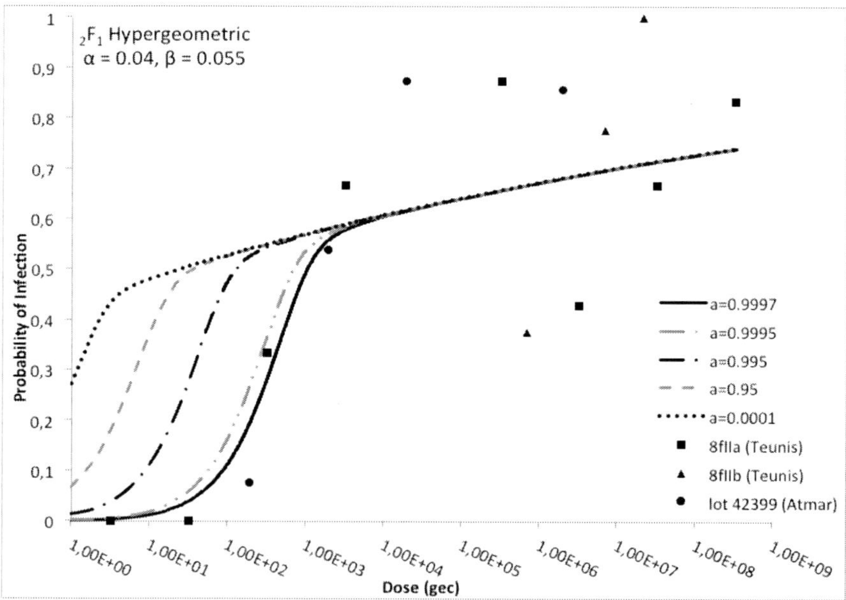

Figure 3. Comparison of the $_2F_1$ Hypergeometric Function at Varying 'a' Values.

The $_2F_1$ dose-response model assuming $\alpha = 0.04$, $\beta = 0.055$, but with different aggregation assumptions demonstrates that the 'a' parameter does impact the dose-response model and the ID_{50} estimate (Figure 3). While the shape of the curves with different 'a' parameters is the same, the ID_{50} shifts with different aggregation assumptions. As the 'a' parameter is decreased toward 0, the ID_{50} value is lowered. The ID_{50} when the virus is aggregated ('a' = 0.9997) is ~1015 gec and when the virus is not aggregated ('a' = 0.0001) the ID_{50} is ~18 gec. This is approximately 2 orders of magnitude difference for the non-aggregated case (a = 0.0001) as compared to the aggregated case (McBride et al. 2013). The same impact on the ID_{50} values is observed with the fractional Poisson model (Figure 4). The fractional Poisson model does not use the 'a' parameter, but uses μ_a, such that when the virus is assumed to be completely disaggregated in the environment then a = 0.0001 and $\mu_a = 1$ virus per aggregate (equation 7). The ID_{50} value when the virus is aggregated ($\mu_a = 579$ viruses per aggregate) is 756 gec and when the dose is disaggregated ($\mu_a = 1$ virus per aggregate) then the ID_{50} value is quite low falling between 1 and 2 gec. Thus, a high degree of aggregation assumed by both the $_2F_1$ and the fractional Poisson yields ID_{50} estimates that are orders of magnitude larger than when no aggregation is assumed. However, questions remain about the reasonableness of these models and their assumptions on aggregation, susceptibility, and immunity (Schmidt 2015). Moreover, the ID_{50} results from zeroing the aggregation parameter after fitting the $_2F_1$ hypergeometric (Figure 3) or the fractional Poisson (Figure 4) lead to results that are not congruent with the published data, which depicts no one being infected at doses <100 gec.

A similar result was observed when risk results initially run with the $_2F_1$ hypergeometric model ($\alpha = 0.04$, $\beta = 0.055$, a = 0.9997) (Teunis et al. 2008) were re-run using $_1F_1$ dose-response parameters estimated from oyster outbreak data (ID_{50} values presented in Table 4) (Thebault et al. 2013). The results found the median probability of infection was several orders of magnitude greater using the outbreak parameters, which did not assume aggregation, as opposed to using the human challenge dose-response parameters, which assumed aggregation (a = 0.9997). Thus, the authors concluded that use of the human challenge dose-response parameters (2F1_TGIa+b) noticeably underestimated the risks of infection (Mok et al. 2014).

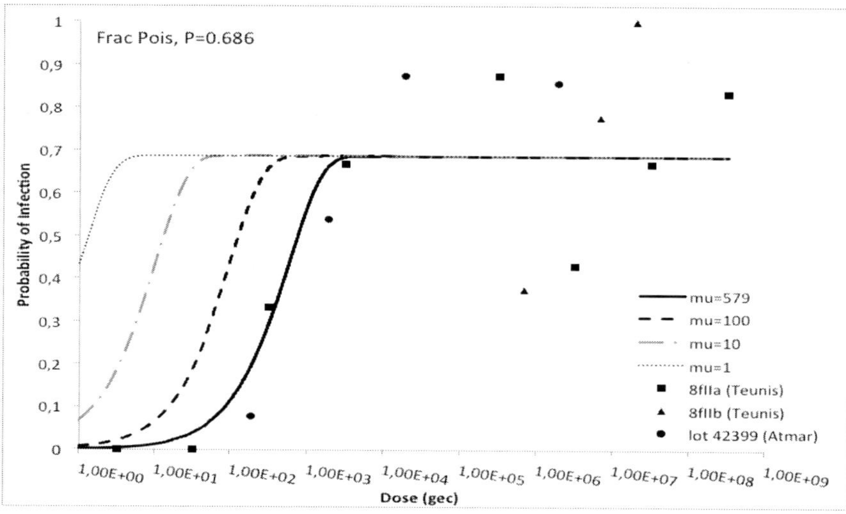

Figure 4. Comparison of the Fractional Poisson Model at Varying μ_a Values.

Discussion of Aggregated Model Assumptions

The beta distribution estimates (α and β) are much less than 1 (Table 3) indicating that the majority of the probability densities for the single hit infectivity (p_m) fall in the extremes of 0 or 1. Aggregation impacts the bathtub shape because it was observed that when the same dose-response model was fit with varying μ_a parameters (from 1 to ~600) then the α and β moved farther below 1 (Schmidt 2015). This characteristic bathtub-shaped distribution can be observed in Figure 5 for 2F1_TGIa and Figure 6 for 2F1_TGIa+b where the majority of the probability densities fall into the extreme values of 0 and 1. This was also observed by others and led to the conclusion that most people are either totally susceptible or totally immune (Messner et al. 2014). However, with the plateauing of the dose-response models (Figures 3 and 4) and numerous data points falling outside of the confidence or credible bounds, perfect immunity or susceptibility appears unlikely. Also of note, the MLE estimated μ_a values for 3 scenarios were quite large (~400 to 1000 viruses per aggregate) and led to corresponding ‘a’ parameters that were very close to 1 (equation 7) indicating strong aggregation. The 2F1_AGI scenario had a small μ_a value (~4 viruses per aggregate), but this ‘a’ value was arbitrarily selected because the procedure failed to converge at higher values presumably due to the small sample size (McBride 2014).

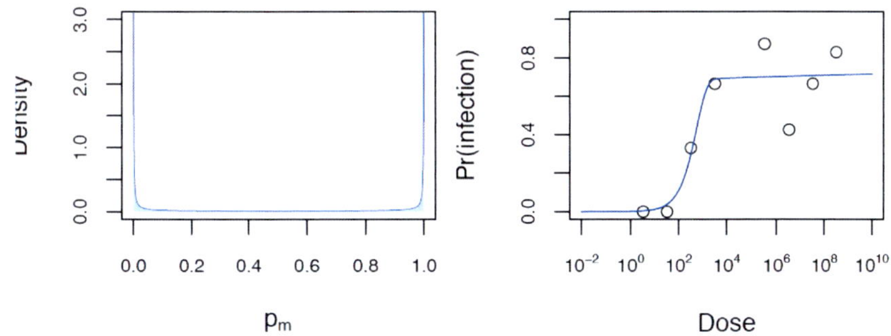

Figure 5. Fitted parameter values to the $_2F_1$ hypergeometric model using 8fIIa data only (2F1_TGIa) are $\alpha = 0.00535$, $\beta = 0.00251$, $a = 0.9998$, (a) The evaluated beta (α, β) density of p_m, (b) the $_2F_1$ dose-response model with the observed data.

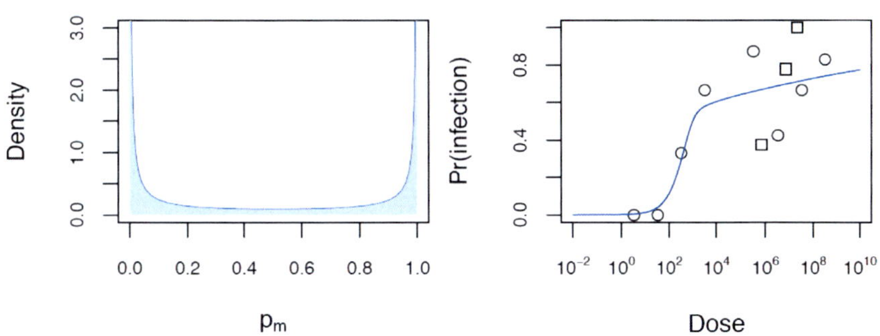

Figure 6. Fitted parameter values to the $_2F_1$ hypergeometric model using 8fIIa+b data only (2F1_TGIa+b) are $\alpha = 0.04$, $\beta = 0.055$, $a = 0.9997$, (a) The evaluated beta (α, β) density of p_m, (b) the $_2F_1$ dose-response model with the observed data.

Choice of Dose-Response Model for Use in QMRA

Norovirus dose-response is not straightforward and many dose-response models are available with numerous assumptions, approximations, and considerations made. When selecting a model for use in QMRA all assumptions, approximations, and considerations of should be clearly reported. Also, variability and uncertainty should be incorporated into dose-response analysis to properly attributing the sources of "scatter", which is a challenge (Haas et al. 2014). Ultimately, it has been proposed that multiple norovirus dose-response models be used to QMRA to describe the uncertainty (Van Abel et al. 2016).

SUMMARY/CONCLUSION

Several unsettled research questions about norovirus remain, which require the statement of explicit assumptions and complicate the performance of QMRAs. Still, QMRAs can be a valuable tool understanding risk of transmission and control of norovirus in a variety of exposure settings. Further QMRAs can be used for the testing of assumptions. Best QMRA practices will accept and account for the uncertainty and variability inherent in both the dose response and exposure assessment in their estimation of risk. Lastly, more data is clearly needed to fully understand norovirus transmission and control. Key data on virus aggregation state and infectivity in environmental matrices is needed and anticipated from recently developed tissue culture methods. Additional human challenge data and better outbreak data is needed for other genotypes of norovirus to better inform dose-response models. In the interim when performing QMRA, assumptions must be clearly stated and deliberately challenged in the risk assessments, and mulitple dose-response models should be considered to adequately capture variabilty and uncertainty.

REFERENCES

Abramowitz M, Stegun IA. 1965. *Handbook of Mathematical Functions.* New York: Dover Publications.

Atmar RL, Neill FH, Romalde JL, Le Guyader F, Woodley CM, Metcalf TG, Estes MK. 1995. Detection of Norwalk virus and hepatitis A virus in shellfish tissues with the PCR. *Appl Environ Microbiol* 61: 3014-3018.

Atmar RL. 2010. *Noroviruses - State of the Art. Food Environ Virol* 2: 117-126.

Atmar RL, Bernstein DI, Harro CD, Al-Ibrahim MS, Chen WH, Ferreira J, Estes MK, Graham DY, Opekun AR, Richardson C, Mendelman PM. 2011. Norovirus vaccine against experimental human Norwalk Virus illness. *N Engl J Med* 365: 2178-2187.

Atmar RL, Opekun AR, Gilger MA, Estes MK, Crawford SE, Neill FH, Ramani S, Hill H, Ferreira J, Graham DY. 2014. Determination of the 50% human infectious dose for Norwalk virus. *J Infect Dis* 209: 1016-1022.

Bambic D, McBride G, Miller W, Stott R, Wuertz S. 2011. *Quantification of Pathogens and Sources of Microbial Indicators for QMRA in Recreational Waters.* Alexandria, VA: WERF (Water Environment Research Federation) and IWA Publishing.

Bartrand TA, Weir MH, Haas CN. 2008. Dose-response models for inhalation of bacillus anthracis spores: Interspecies comparisons. *Risk Anal* 28: 1115-1124.

Barker SF, O'Toole J, Sinclair MI, Leder K, Malawaraarachchi M, Hamilton AJ. 2013a. A probabilistic model of norovirus disease burden associated with greywater irrigation of home-produced lettuce in Melbourne, Australia. *Water Res* 47: 1421-1432.

Barker SF, Packer M, Scales PJ, Gray S, Snape I, Hamilton AJ. 2013b. Pathogen reduction requirements for direct potable reuse in Antarctica: evaluating human health risks in small communities. *Sci Total Environ* 461-462: 723-733.

Bernstein DI, Atmar RL, Lyon GM, Treanor JJ, Chen WH, Jiang X, Vinjé J, Gregoricus N, Frenck RW, Moe CL, Al-Ibrahim MS, Barrett J, Ferreira J, Estes MK, Graham DY, Goodwin R, Borkowski A, Clemens R, Mendelman PM. 2015. Norovirus vaccine against experimental human GII.4 virus illness: a challenge study in healthy adults. *J Infect Dis* 211: 870-878.

Carlsson B, Kindberg E, Buesa J, Rydell GE, Lidon MF, Montava R, Mallouh RA, Grahn A, Rodriguez-Diaz J, Bellido J, Arnedo A, Larson G, Svensson L. 2009. The G428A nonsense mutation in FUT2 provides strong but not absolute protection against symptomatic GII.4 norovirus infection. *PLoS One* 4: e5593.

Costantini V, Loisy F, Joens L, Le Guyader FS, Saif LJ. 2006. Human and animal enteric caliciviruses in oysters from different coastal regions of the United States. *Appl Environ Microbiol* 72: 1800-1809.

Currier RL, Payne DC, Staat MA, Selvarangan R, Shirley SH, Halasa N, Boom JA, Englund JA, Szilagyi PG, Harrison CJ, Klein EJ, Weinberg GA, Wikswo ME, Parashar U, Vinjé J, Morrow AL. 2015. Innate susceptibility to norovirus infections influenced by FUT2 genotype in a United States pediatric population. *Clin Infect Dis* 60: 1631-1638.

Debbink K. 2015. Ancestry-based differences in norovirus susceptibility: Implications for understanding global infection patterns. *Clin Infect Dis* 60: 1639-1641.

de Graaf M, van Beek J, Vennema H, Podkolzin AT, Hewitt J, Bucardo F, Templeton K, Mans J, Nordgren J, Reuter G, Lynch M, Rasmussen LD, Iritani N, Chan MC, Martella V, Ambert-Balay K, Vinjé J, White PA, Koopmans MP. 2015. Emergence of a novel GII.17 norovirus – End of the GII.4 era? *Euro Surveill* 20: pii 21178.

DePaola A, Jones JL, Woods J, Burkhardt W 3rd, Calci KR, Krantz JA, Bowers JC, Kasturi K, Byars RH, Jacobs E, Williams-Hill D, Nabe K. 2010. Bacterial and viral pathogens in live oysters: 2007 United States market survey. *Appl Environ Microbiol.* 76: 2754-2768.

Ettayebi K, Crawford SE, Murakami K, Broughman JR, Karandikar U, Tenge VR, Neill FH, Blutt SE, Zeng XL, Qu L, Kou B, Opekun AR, Burrin D, Graham DY, Ramani S, Atmar RL, Estes MK. 2016. Replication of human noroviruses in stem cell-derived human enteroids. *Science* 353: 1387-1393.

Ferrer-Admetlla A, Sikora M, Laayouni H, Esteve A, Roubinet F, Blancher A, Calafell F, Bertranpetit J, Casals F. 2009. A natural history of FUT2 polymorphism in humans. *Mol Biol Evol* 26:19 93-2003.

Floyd R, Sharp DG. 1977. Aggregation of poliovirus and reovirus by dilution in water. *Appl Environ Microbiol* 33: 159-167.

Frenck R, Bernstein DI, Xia M, Huang P, Zhong W, Parker S, Dickey M, McNeal M, Jiang X. 2012. Predicting Susceptibility to Norovirus GII.4 by Use of a Challenge Model Involving Humans. *J Infect Dis* 206: 1386-1393.

Furumoto WA, Mickey R. 1967. A Mathematical Model for Infectivity-Dilution Curve of Tobacco Mosaic Virus - Theoretical Considerations. *Virology* 32: 216-223.

Girones R, Ferrus MA, Alonso JL, Rodriguez-Manzano J, Calgua B, Correa Ade A, Hundesa A, Carratala A, Bofill-Mas S. 2010. Molecular detection of pathogens in water--the pros and cons of molecular techniques. *Water Res* 44: 4325-4339.

Goodridge L, Goodridge C, Wu J, Griffiths M, Pawliszyn J. 2004. Isoelectric point determination of norovirus virus-like particles by capillary isoelectric focusing with whole column imaging detection. *Anal Chem* 76: 48-52.

Graham DY, Jiang X, Tanaka T, Opekun AR, Madore HP, Estes MK. 1994. Norwalk virus infection of volunteers: new insights based on improved assays. *J Infect Dis* 170: 34-43.

Greening G, Lake R, Hudson A, Cressey P. 2009. *Risk Profile: Norovirus in Mollusca* (Raw). Christchurch, NZ: Institute of Environmental Science and Research, Ltd.

Ha S, Choi IS, Choi C, Myoung J. 2016. Infection models of human norovirus: challenges and recent progress. *Arch Virol* 161: 779-788.

Haas CN, Rose JB, Gerba C, Regli S. 1993. Risk assessment of virus in drinking water. *Risk Anal* 13: 545-552.

Haas CN, Rose JB, Gerba CP. 2014. *Quantitative Microbial Risk Assessment,* 2nd Edition. New York: John Wiley and Sons, Inc.

Han J, Ji L, Shen Y, Wu X, Xu D, Chen L. 2015. Emergence and predominance of norovirus GII.17 in Huzhou, China, 2014-2015. *Virol J* 12: 139.

Hankin RKS Package 'gsl' for R software. 2013.

Harrington PR, Lindesmith L, Yount B, Moe CL, Baric RS. 2002. Binding of Norwalk virus-like particles to ABH histo-blood group antigens is blocked by antisera from infected human volunteers or experimentally vaccinated mice. *J Virol* 76: 12335-12343.

Hunter PR, Payment P, Ashbolt NJ, Bartram J. 2003. Chapter 3: *Assessment of Risk, Assessing microbial safety of drinking water: Improving approaches and methods.* Padstow, Cornwall, UK: TJ International (Ltd); pp. 79-109.

Hutson AM, Atmar RL, Graham DY, Estes MK. 2002. Norwalk virus infection and disease is associated with ABO histo-blood group type. *J Infect Dis* 185: 1335-1337.

Johnson FB, Bodily AS. 1975. Effect of environmental pH on adenovirus-associated virus. *Proc Soc Exp Biol Med* 150: 585-590.

Jones MK, Grau KR, Costantini V, Kolawole AO, de Graaf M, Freiden P, Graves CL, Koopmans M, Wallet SM, Tibbetts SA, Schultz-Cherry S, Wobus CE, Vinjé J, Karst SM. 2015. Human norovirus culture in B cells. *Nat Protoc* 10: 1939-1947.

Kelly RJ, Rouquier S, Giorgi D, Lennon GG, Lowe JB. 1995. Sequence and expression of a candidate for the human secretor blood group alpha(1,2)fucosyltransferase gene (FUT2). Homozygosity for an enzyme-inactivating nonsense mutation commonly correlates with the non-secretor phenotype. *J Biol Chem* 270: 4640-4649.

Koepf W. 1998. *Hypergeometric Summation: An Algorithmic Approach to Summation and Special Function Identities*. Germany: Vieweg Verlag.

Langlet J, Gaboriaud F, Gantzer C. 2007. Effects of pH on plaque forming unit counts and aggregation of MS2 bacteriophage. *J Appl Microbiol* 103: 1632-1638.

Langlet J, Gaboriaud F, Duval JFL, Gantzer C. 2008. Aggregation and surface properties of F-specific RNA phages: Implication for membrane filtration processes. *Water Res* 42: 2769-2777.

Le Guyader FS, Neill FH, Dubois E, Bon F, Loisy F, Kohli E, Pommepuy M, Atmar RL. 2003. A semiquantitative approach to estimate Norwalk-like virus contamination of oysters implicated in an outbreak. *Int J Food Microbiol* 87: 107-112.

Le Guyader FS, Bon F, DeMedici D, Parnaudeau S, Bertone A, Crudeli S, Doyle A, Zidane M, Suffredini E, Kohli E, Maddalo F, Monini M, Gallay A, Pommepuy M, Pothier P, Ruggeri FM. 2006. Detection of multiple noroviruses associated with an international gastroenteritis outbreak linked to oyster consumption. *J Clin Microbiol* 44: 3878-3882.

Le Guyader FS, Le Saux JC, Ambert-Balay K, Krol J, Serais O, Parnaudeau S, Giraudon H, Delmas G, Pommepuy M, Pothier P, Atmar RL. 2008. Aichi Virus, Norovirus, Astrovirus, Enterovirus, and Rotavirus Involved in Clinical Cases from a French Oyster-Related Gastroenteritis Outbreak. *J Clin Microbiol* 46: 4011-4017.

Le Guyader FS, Krol J, Ambert-Balay K, Ruvoen-Clouet N, Desaubliaux B, Parnaudeau S, Le Saux JC, Ponge A, Pothier P, Atmar RL, Le Pendu J. 2010. Comprehensive analysis of a norovirus-associated gastroenteritis outbreak, from the environment to the consumer. *J Clin Microbiol* 48: 915-920.

Leon JS, Kingsley DH, Montes JS, Richards GP, Lyon GM, Abdulhafid GM, Seitz SR, Fernandez ML, Teunis PF, Flick GJ, Moe CL. 2011. Randomized, double-blinded clinical trial for human norovirus inactivation in oysters by high hydrostatic pressure processing. *Appl Environ Microbiol* 77: 5476-5482.

Lindesmith L, Moe C, Marionneau S, Ruvoen N, Jiang X, Lindbland L, Stewart P, Le Pendu J, Baric R. 2003. Human susceptibility and resistance to Norwalk virus infection. *Nat Med* 9: 548-553.

Liu P, Hsiao HM, Jaykus LA, Moe C. 2010. Quantification of Norwalk virus inocula: Comparison of endpoint titration and real-time reverse transcription-PCR methods. *J Med Virol* 82: 1612-1616.

Lopman BA, Trivedi T, Vicuna Y, Costantini V, Collins N, Gregoricus N, Parashar U, Sandoval C, Broncano N, Vaca M, Chico ME, Vinjé J, Cooper PJ. 2015. Norovirus infection and disease in an Ecuadorian birth cohort: association of certain norovirus genotypes with host FUT2 secretor status. *J Infect Dis* 211: 1813-1821.

Mans J, Murray TY, Nadan S, Netshikweta R, Page NA, Taylor MB. 2016. Norovirus diversity in children with gastroenteritis in South Africa from 2009 to 2013: GII.4 variants and recombinant strains predominate. *Epidemiol Infect* 144: 907-916.

Marionneau S, Ruvoen N, Le Moullac-Vaidye B, Clement M, Cailleau-Thomas A, Ruiz-Palacois G, Huang PW, Jiang X, Le Pendu J. 2002. Norwalk virus binds to histo-blood group antigens present on gastroduodenal epithelial cells of secretor individuals. *Gastroenterology* 122: 1967-1977.

Marionneau S, Airaud F, Bovin NV, Le Pendu J, Ruvoen-Clouet N. 2005. Influence of the combined ABO, FUT2, and FUT3 polymorphism on susceptibility to Norwalk virus attachment. *J Infect Dis* 192: 1071-1077.

McBride GB, Stott R, Miller W, Bambic D, Wuertz S. 2013. Discharge-based QMRA for estimation of public health risks from exposure to stormwater-borne pathogens in recreational waters in the United States. *Water Res* 47: 5282-5297.

McBride GB. (ed.) 2014. Norovirus dose-response in sewage-related QMRA: The importance of virus aggregation. 7th International Congress on Environmental Modelling and Software. San Diego, California, USA.

Mesquita JR, Barclay L, Nascimento MS, Vinjé J. 2010. Novel norovirus in dogs with diarrhea. *Emerg Infect Dis* 16: 980-982.

Messner MJ, Berger P, Nappier SP. 2014. Fractional Poisson-A Simple Dose-Response Model for Human Norovirus. *Risk Anal* 34: 1820-1829.

Michen B, Graule T. 2010. Isoelectric points of viruses. *J Appl Microbiol* 109: 388-397.

Mok H-F, Barker S, Hamilton A. 2014. A probabilistic quantitative microbial risk assessment model of norovirus disease burden from wastewater irrigation of vegetables in Shepparton, Australia. *Water Res* 54: 347-362.

Nordgren J, Kindberg E, Lindgre PE, Matussek A, Svensson L. 2010. Norovirus gastroenteritis outbreak with a secretor-independent susceptibility pattern, Sweden. *Emerg Infect Dis* 16: 81-87.

Overby LR, Barlow GH, Doi RH, Jacob M, Spiegelman S. 1966. Comparison of two serologically distinct ribonucleic acid bacteriophages, II. Properties of the nucleic acids and coat proteins. *J Bacteriol* 92: 739-745.

Parra GI, Green KY. 2015. Genome of emerging Norovirus GII.17, United States, 2014. *Emerg Infect Dis* 21: 1477-1479.

Payne DC, Parashar UD, Lopman BA. 2015. Developments in understanding acquired immunity and innate susceptibility to norovirus and rotavirus gastroenteritis in children. *Curr Opin Pediatr* 27: 105-109.

Pearson CE. 1983. *Handbook of Applied Mathematics.* 2nd ed. New York: Van Nostrand Reinhold Company.

Redman JA, Grant SB, Olson TM. 1997. Filtration of recombinant Norwalk virus particles and bacteriophage MS2 in quartz sand: importance of electrostatic interactions. *Environ Sci Technol* 31: 3378-3383.

Saito M, Goel-Apaza S, Espetia S, Velasquez D, Cabrera L, Loli S, Crabtree JE, Black RE, Kosek M, Checkley W, Zimic M, Bern C, Cama V, Gilman RH. 2013. Multiple Norovirus infections in a birth cohort in a Peruvian periurban community. *Clin Infect Dis* 58: 483-491.

Sakon N, Yamazaki K, Nakata K, Kanbayashi D, Yoda T, Mantani M, Kase T, Takahashi K, Komano J. 2014. Impact of genotype-specific herd immunity on the circulatory dynamism of Norovirus: A 10-year longitudinal study of viral acute gastroenteritis. *J Infect Dis* 211: 879-888.

Sanchez MA, Corcostégui SP, De Broucker CA, Cabre O, Watier-Grillot S, Perelle S, Ambert-Balay K, de Santi VP. 2016. Norovirus GII.17 outbreak linked to an infected post-symptomatic food worker in a french military unit located in France. *Food Environ Virol.* doi: 10.1007/s12560-016-9274-0.

Schijven JF, Teunis PPM, Rutjes SA, Bouwknegt M, Husman AMD. 2011. QMRAspot: A tool for Quantitative Microbial Risk Assessment from surface water to potable water. *Water Res* 45: 5564-5576.

Schmidt PJ. 2015. Norovirus dose–response: are currently available data informative enough to determine how susceptible humans are to infection from a single virus? *Risk Anal* 35: 1364-1383.

Schoen ME, Ashbolt NJ. 2010. Assessing pathogen risk to swimmers at non-sewage impacted recreational beaches. *Environ Sci Technol* 44: 2286-2291.

Seitz SR, Leon JS, Schwab KJ, Lyon GM, Dowd M, McDaniels M, Abdulhafid G, Fernandez ML, Lindesmith LC, Baric RS, Moe CL. 2011. Norovirus infectivity in humans and persistence in water. *Appl Environ Microbiol* 77: 6884-6888.

Shirato H, Ogawa S, Ito H, Sato T, Kameyama A, Narimatsu H, Xiaofan Z, Miyamura T, Wakita T, Ishii K, Takeda N. 2008. Noroviruses distinguish between type 1 and type 2 histo-blood group antigens for binding. *J Virol* 82: 10756-10767.

Shirato H. 2011. Norovirus and Histo-Blood Group Antigens. *Jpn J Infect Dis* 64: 95-103.

Siebenga JJ, Vennema H, Zheng DP, Vinjé J, Lee BE, Pang XL, Ho EC, Lim W, Choudekar A, Broor S, Halperin T, Rasool NB, Hewitt J, Greening GE, Jin M, Duan ZJ, Lucero Y, O'Ryan M, Hoehne M, Schreier E, Ratcliff RM, White PA, Iritani N, Reuter G, Koopmans M. 2009. Norovirus illness is a global problem: emergence and spread of norovirus GII.4 variants, 2001-2007. *J Infect Dis* 200: 802-812.

Silva LD, Bandeira RD, Junior EC, Lima IC, da Penha Júnior ET, Teixeira DM, Siqueira JA, Resque HR, de Abreu Campos EM, Justino MC, Linhares AC, Gabbay YB. 2016. Detection and genetic characterization of the emergent GII.17_2014 norovirus genotype among children with gastroenteritis from Northern Brazil. *Infect Genet Evol* 48: 1-3.

Tan M, Jin M, Xie H, Duan Z, Jiang X, Fang Z. 2008. Outbreak studies of a GII-3 and a GII-4 norovirus revealed an association between HBGA phenotypes and viral infection. *J Med Virol* 80: 1296-1301.

Tan M, Jiang X. 2014. Histo-blood group antigens: a common niche for norovirus and rotavirus. *Expert Rev Mol Med* 16: e5.

Teunis PF, Havelaar AH. 2000. The Beta Poisson dose-response model is not a single-hit model. *Risk Anal* 20: 513-520.

Teunis PF, Moe CL, Liu P, Miller SE, Lindesmith L, Baric RS, Le Pendu J, Calderon RL. 2008. Norwalk virus: how infectious is it? *J Med Virol* 80: 1468-1476.

Teunis PF, Sukhrie FHA, Vennema H, Bogerman J, Beersma MFC, Koopmans MPG. 2015. Shedding of norovirus in symptomatic and asymptomatic infections. *Epidemiol Infect* 143: 1710-1717.

Thebault A, Teunis PF, Le Pendu J, Le Guyader FS, Denis JB. 2013. Infectivity of GI and GII noroviruses established from oyster related outbreaks. *Epidemics* 5: 98-110.

Topping JR, Schnerr H, Haines J, Scott M, Carter MJ, Willcocks MM, Bellamy K, Brown DW, Gray JJ, Gallimore CI, Knight AI. 2009. Temperature inactivation of Feline calicivirus vaccine strain FCV F-9 in comparison with human noroviruses using an RNA exposure assay and reverse transcribed quantitative real-time polymerase chain reaction: A novel method for predicting virus infectivity. *J Virol Methods* 156: 89-95.

Van Abel N, Schoen ME, Kissel J, Meschke JS. 2016. Comparison of risk predicted by multiple norovirus dose-response models and implications for quantitative microbial risk assessment. *Risk Anal.* doi: 10.1111/risa.12616.

Van Trang N, Vu HT, Le NT, Huang P, Jiang X, Anh DD. 2014. Association between norovirus and rotavirus infection and histo-blood group antigen types in Vietnamese children. *J Clin Mircobiol* 52: 1366-1374.

Weisstein EW. 2014. *Calculus and Analysis: Hypergeometric Function* [Web Resource]: From MathWorld-A Wolfram Web Resources.

Wong K., Mukherjee B., Kahler A.M., Zepp R., Molina M. 2012. Influence of Inorganic Ions on Aggregation and Adsorption Behaviors of Human Adenovirus. *Environ Sci & Technol* 46: 11145-53.

WHO (World Health Organization). 2011. *Guidelines for Drinking-water Quality.* 4th Edition ed. Geneva, Switzerland: WHO Press.

In: Noroviruses
Editor: Jesús L. Romalde

ISBN: 978-1-53611-939-8

Chapter 6

SOLAR RADIATION AND PHOTO-FENTON SYSTEMS: NOVEL APPLICATIONS FOR NOROVIRUS DISINFECTION IN WATER

Enrique Rivadulla[1], Irene García-Fernández[2], Jesús L. Romalde[1,*], Pilar Fernández-Ibáñez[2,†] and David Polo[1,‡]

[1]Departamento de Microbiología y Parasitología, CIBUS-Facultad de Biología, Universidade de Santiago de Compostela, Santiago de Compostela, Spain

[2]Plataforma Solar de Almería, CIEMAT, Tabernas, Almería, Spain

* Corresponding Author address: Departamento de Microbiología y Parasitología, CIBUS-Facultad de Biología. Universidade de Santiago de Compostela. Campus Vida s/n. 15782, Santiago de Compostela. Spain; Email: jesus.romalde@usc.es.

† Correspondng Author Address School of Engineering. Ulster University. Shore Rd. Newtownabbey. BT37 0QB. Northen Ireland.

‡ Corresponding Author address: Departamento de Microbiología y Parasitología, CIBUS-Facultad de Biología. Universidade de Santiago de Compostela. Campus Vida s/n. 15782, Santiago de Compostela. Spain; Email: d.polo.montero@gmail.com

Abstract

Ensuring the safety and quality of drinking water is essential in the prevention of waterborne diseases, especially in developing countries where water scarcity and lack of access to sanitation is common. Noroviruses (NoV) are the leading cause of non-bacterial acute gastroenteritis worldwide, being the second most common cause of diarrheal death in children under 5 years of age. Through the review of previous published works together with original research, this chapter focus on the feasibility of alternative water treatment processes, such as solar disinfection (SODIS), Fenton and photo-Fenton systems, for the virus inactivation.

Negligible viral reduction was achieved in Fenton systems, and only partial disinfection was achieved under SODIS conditions. On the other hand, photo-Fenton process significantly increased the disinfection power. A synergistic effect between Fe^{2+}/H_2O_2 and sunlight leads to significant disinfection rates of murine norovirus (employed as surrogate of human norovirus) at middle environmental temperatures and without an intensive energy-light source requirement. Nevertheless, results presented here suggest solar photo-Fenton technology as a feasible alternative treatment for viral disinfection, in both high and low -income settings, to reach acceptable water quality for human consumption or wastewater reuse.

Keywords: norovirus, murine norovirus, SODIS, phto-Fenton

Introduction

Water scarcity and the lack of access to sanitation in developing countries is one of the main problems for the human populations (Polo et al., 2015). Surface waters receive the impact of urban or community wastewater contaminated with fecal material. As a consequence, a quarter of global population is consuming fecally contaminated water (WHO - UNICEF, 2014), which can constitute a dangerous route of transmission for pathogenic bacteria, protozoa and virus, causing of a variety of human diseases as well as an important global health impact (UNICEF, 2012).

Conventional sewage and water treatment methods, even when properly functioning, have been proven ineffective for the complete removal and inactivation of some human pathogens (Haas et al., 1993). In fact, contaminated water constitutes one of the most common routes for enteric waterborne diseases (Fewtrell and Bartram, 2001), directly by drinking untreated or improperly treated water and indirectly through contamination of food by irrigation with polluted or faecal contaminated water (Maunula et al., 2012). The control of the performance of conventional treatment methods is routinely performed by the analysis of fecal coliforms, but viral contamination assessment is scarce or nule because it is not feasible in a cost-efficient and timely manner (Gall et al., 2015).

Household water treatment and storage (HWTS) and advanced oxidation processes (AOPs) can contribute to global efforts in order to provide universal access to safe water. These methods have been demonstrated as a useful tool to reduce the incidence of waterborne diseases, mostly those with bacterial or parasitic etiology, in regions with difficult or inexistent access to treated drinking water. In addition, they constitute also cheap, easy to use and sustainable water treatments that could achieve basic criteria for a fast installation in this kind of population areas.

Diarrheal disease constitutes the second most common cause of morbidity overall in children under the age of 5 years. Each year, 1.9 million of children under 5 years die due to diarrheal disease, representing approximately 19% of all deaths in developing countries (Boschi-Pinto et al., 2008). Improving water quality, sanitation and hygiene could considerably reduce this percentage (Prüss-Üstün et al., 2008).

In the last decade, and after the introduction of rotavirus vaccination programs, noroviruses (NoV) became the leading etiological agents of foodborne disease in all age groups (Rodriguez-Lazaro et al., 2012; Scallan et al., 2011; Atmar and Estes, 2001; Koopmans and Duizer, 2004). A large diversity of NoV strains has been described on gastroenteritis outbreaks and linked to transmission through drinking water (Ayukekbong et al., 2015). Furthermore NoV possess a high environmental stability, a low infection dose, are persistent in water supply systems and resistant to common disinfection process (Vaughan et al., 2014; Bosch, 2010). NoV belong to the family *Caliciviridae*, are a group of non-enveloped, icosahedral viruses with approximately 7.5 kb positive-sense ssRNA genome. NoV are genetically diverse and are composed of at least 41 genotypes, classified into seven genogroups (GI-GVII) (Vinjé, 2015), of which genogroups GI, GII and GIV infect humans (Zheng et al., 2006).

The prototype of NoV strain was first described in 1972 using immune electron microscopy on stool of volunteers inoculated with fecal filtrates from a group of elementary who had suffered an outbreak of gastroenteritis (Kapikian et al., 1972). In a contemporary survey Atmar et al., (2014) could observe, in an experiment with inoculated volunteers, viral shedding peaks after 4 days of virus infection. The volunteers of the experiment began shedding after 8h of the virus inoculation and the highest concentration of shedding occurred after resolution of symptoms in the 69% of the cases. Studies have shown that NoV can spread easily, as few as 18-2800 viral particles can cause infection (Barreira et al., 2010; Teunis et al., 2008), and on the other side a gram of stool from an infected host contain greater than $1x10^{10}$ RNA copies (Chan et al., 2006). This evidences, added to the environmental stability, resistance to disinfectants and thermal stability of NoV, produce that a single infected person has the potential to infect thousands of others (Pringle et al., 2015).

Despite the fact that NoV reliable burden is difficult to establish, not only for the reasons previously mentioned but for the unavailability of simple clinical diagnostic tests (Patel et al., 2008), it has been associated to approximately 18% of diarrheal disease among human population and is the second most common cause of diarrheal death in children under 5 years of age. Also NoV has been established as the most common diarrheal cause in humans over the age of 5 years (Lopman and Grassly, 2016), with healthcare associated economic losses of up to £115 million ($184 million) in high-income settings (Head and Lopman, 2016). In developing countries, NoV are responsable for over 218.000 deaths annually (Kocher and Yuan, 2015) and the severe disease burden afects the very young (Baehner et al., 2016).

Given the morbidity, mortality and economic costs exposed, there is a need of control about NoV infection. Latest studies, like the recent discovery about NoV replication in stem cell–derived human enteroids under specific lab conditions (Ettayebi et al., 2016) or the surveys about vaccines against NoV that are being developed (Pringle et al., 2015; Kocher and Yuan 2015), have improved our understanding about this human pathogen. However, there are several challenges, such as the incomplete understanding of immunity to this virus or the fact that NoV cultivation is an expensive and time expending methodology. Nevertheless, murine norovirus (MNV) has been demonstrated as a useful alternative because their similarity in genetic and environmental properties (Bae and Schwab, 2008).

The main objective of this chapter is to compare the effectiveness of Solar disinfection (SODIS), as HWTS method, and other AOPs like homogeneous Fenton and solar photo-Fenton systems for the disinfection of MNV (as human NoV surrogate) in water under field conditions.

SODIS

SODIS is a useful HWTS method that became popular in recent years. SODIS technique consists in exposition of contaminated water (volumes up to 3 l) with relatively low turbidity (<30 NTU) in transparent containers (reactors) that can be of glass or plastic (usually polyethylene-terephthalate [PET] bottles) to direct sunlight for at least 6h, during the maximum intensity of radiation or for two consecutive days if it is cloud cover. The water must be consumed in the next 24 h after SODIS disinfection, which is the recommended period to avoid re-growth of post-exposure pathogens (McGuigan et al., 2012).

The biocidal effects of this technique are attributed to UV portion of solar radiation and solar heating (Boyle et al., 2008; Jagger, 1981). There are three types of ultraviolet radiation (UV), UV-C (λ = 100-280 nm), UV-B (λ = 280-320 nm) and UV-A (λ = 320-400) and only UV-A and part of UV-B reach the earth surface (Goodsel, 2001). UV-A can produce indirect damage to DNA and structural components of cells through photosensitizers and may generate in water reactive oxygen species (ROS) like singlet oxygen ($^1O_2$), superoxide (O_2^-), hydrogen peroxide (H_2O_2), and hydroxyl radicals ($^\bullet OH$). These oxygen compounds could cause single strand breaks, modifications on nucleic bases, as well as induction of protein and lipid membrane oxidations (Reed et al., 2000; Eisenstark, 1987; Jagger, 1981), which lead to serious damage to bacteria, protozoa, fungi and virus. Among these pathogens, viruses seem to be the most resistant entities to SODIS (McGuigan et al., 2012).

In order to determine the efficacy of the method for virus elimination, Polo et al., (2015) performed SODIS experiments with MNV as a surrogate of human NoV at the Plataforma Solar de Almería, Tabernas dessert (Spain). Experiences were carried out in PET reactors (330 ml) with distilled water contaminated with an initial concentration of 10^3 plaque forming units per ml (pfu ml^{-1}) of the virus, that were exposed directly to the action of natural solar radiation during 2, 4, 6 and 8h.

A solar energy unit, Q_{UV}, and UV-A radiometer (295–385 nm, Model CUV3, Kipp & Zonen, Netherlands) were used in order to measure solar radiant energy rate incident on a surface per unit area (W m^{-2}) and to take a plotted of inactivation kinetics.

UV dose (kJ m^{-2}) is given by the Equation:

$$dose = I \times \Delta t$$

where I is the average irradiation intensity, W/m^2, and Δt is the experimental time, in seconds.

The plotted of inactivation kinetics is measured as function of cumulative energy per unit of volume (Q_{UV}, kJ l^{-1}) received by the bottles, and calculated by Equation:

$$Q_{UV,n} = Q_{UV,n-1} + \frac{\Delta t_n \overline{UV}_{G,n} A_r}{V_t} \qquad \Delta t_n = t_n - t_{n-1}$$

where $Q_{uv,n}$, $Q_{uv,\,n-1}$, are the UV energy accumulated per unit volume (kJ l^{-1}) at times n and n-1, respectively, $UV_{G,n}$ is the average incident irradiation on the irradiated area, Δt_n is the experimental time of sample, A_r is the illuminated area of the solar bottle (m^2), and V_t is the total volume (l) of treated water.

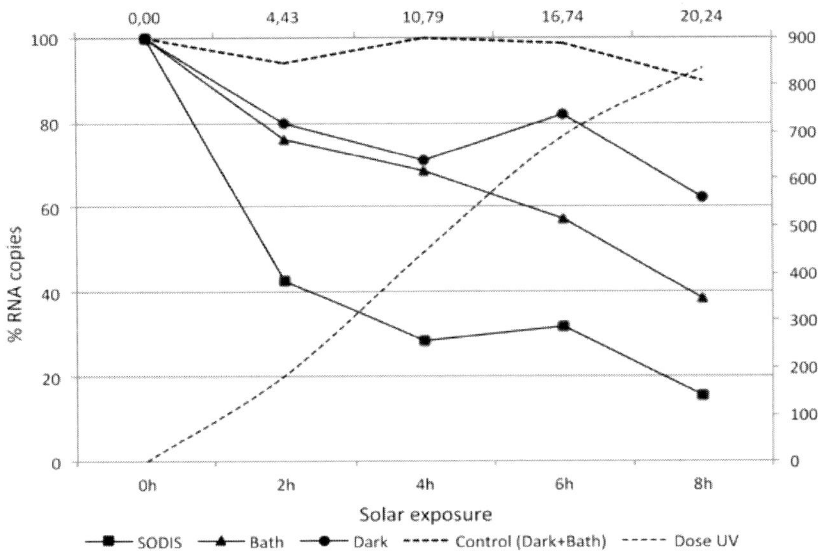

Figure 1. MNV RNA removal expressed as % RNA copies 100 ml^{-1} along the exposure time (2, 4, 6 and 8h). Solar UV dose (W m^{-2}) and cumulative UV radiation (Q_{UV}, kJ l^{-1}) are also shown in secondary axis. (Adapted from Polo et al., 2015).

To discriminate the effect of solar radiation parallel assays under controlled temperatures and in darkness conditions were performed. Reactor bottles under controlled temperatures (Bath-bottles) were maintained at 25 ± 1°C in a cooled water bath, and reactor bottles in darkness (Dark-bottles) were wrapped in opaque aluminium foil and exposed to solar radiation. Moreover Control-bottles were wrapped in opaque aluminium foil and maintained at 25 ± 1°C in a cooled water bath.

Maximum local noon UV irradiances recorded were 38.2 Wm^{-2}. The maximum accumulated UV dose and Q_{UV} at the end of the exposure period were 834.876 kJ m^{-2} and 20.24 kJ l^{-1} . Maximum water temperatures reached were 40.7°C, SODIS-bottles, 39.7°C in Dark-bottles, 30.9°C in Bath-bottles and 27.8°C in Control-bottles. Under these conditions, reductions of MNV RNA levels were observed, being greater those in SODIS reactors (Figure 1).

Table 1. Quantification data of MNV expressed as viral RNA copies 100 ml^{-1} for each solar exposure condition along the study period (t). Each data is the geometric mean of three replicates. Standard deviation is shown in brackets

MNV-1	Log RNA copies/100 ml^{-1}			
t	SODIS	Bath	Dark	Control
0h	4.77 ^{A,a}	4.77 ^{A,a}	4.77 ^{A,a}	4.77 ^{A,a}
2h	4.40 ^{AB,a}	4.65 ^{AB,a}	4.67 ^{AB,a}	4.74 ^{A,a}
4h	4.22 ^{AB,b}	4.60 ^{B,ab}	4.62 ^{B,ab}	4.77 ^{A,a}
6h	4.27 ^{AB,c}	4.52 ^{BC,bc}	4.68 ^{B,ab}	4.76 ^{A,a}
8h	3.96 ^{B,d}	4.35 ^{C,c}	4.56 ^{B,b}	4.72 ^{A,a}
r*	0.81	0.42	0.20	0.05

*, Reduction of viral charge expressed in log-units. Statistical differences among results of the different periods within the same treatment are indicated by different capital letters. Statistical differences among results of the different treatments within the same sampling time are indicated by different small letters. Results with the same letter did not show significant differences ($p > 0.05$)

The average viral quantification, made by RT-qPCR, at initial time (0 h of solar exposure) was 5.91(±0.59)×10^4 RNA copies 100 ml^{-1} (RNAc 100 ml^{-1}). After 8h of solar exposure, average quantification values in SODIS bottles were 9.24(±3.91)×10^3 RNAc 100 ml^{-1}. These values represent an average decrease of 84.4% (0.81 log) for MNV (Table 1). The average decreases for Bath-, Dark- and Control-bottles were 61.6% (0.42 log), 37.4% (0.20 log) and 10.3% (0.05 log) respectively (Table 1). Although RNA copies remained high (~10^3 RNAc 100 ml^{-1}) results showed a reduction of 0.81 log units under SODIS conditions after 8h, that means an important reduction in RNA levels.

With regard to the inactivation kinetic for MNV, a clear stabilization of the inactivating effect after 4 h was observed. This effect could not be related in a direct way to solar exposure since values of solar exposure at 6 h were still quite high. Therefore, factors such as viral aggregation could affect the inactivation kinetics through subpopulations of non-easily-inactivated viruses.

Infectivity assays carried out in the same survey, inoculating water on RAW (mouse monocyte macrophage) cell-line, reflected that water samples from control-bottles and Dark-bottles maintained their infectivity capacity. Bath-bottles showed a one third decrease (33.4%) in their infectivity capacity after 8h. Reactors with SODIS conditions showed a decrease in the infectivity capacity after 6 h (33.4%) and 8 h (66.7%) in MNV trials. In addition, a delay in the appearance of the cytopathic effect (CPE) from 2-3 days to 6 days in water samples from Bath-bottles (after 8 h of exposure) and SODIS-bottles (after 4, 6 and 8 h of exposure) was observed. In a similar experience Harding and Schwab (2012) reported 0.4 log reduction after 2.5 hour and 1.4 log reduction after 6h in infectious MNV by plaque assay and determined by inoculed the samples in RAW cell cultures.

As mentioned above, SODIS principles rely on solar UV radiation and water temperature. The comparison between exposure conditions suggests a greater importance of sunlight radiation over the temperature as a principal factor of viral reduction. Previous studies have reported that high temperatures have a major effect on viral capsid proteins but limited effect on viral genome (Hewitt et al., 2009; Baert et al., 2008; Nuanualsuwan and Cliver, 2003). Viral inactivation by heat relies, therefore, in the loss of ability to bind with its host cell due to structural changes in the viral capsid proteins that disrupt the specific structures needed to recognize and bind the host cells. Besides these facts, the natural way of indirect transmission for enteric viruses like NoV confers high stability in hard environments outside host's body.

In addition, there is a synergetic effect between heating and UV inactivation over 45ºC, which improve disinfection (McGuigan et al., 1998; Wegelin et al., 1994). Although these temperatures were never reached in the study of Polo and coworkers (2015), the best results were also achieved with the combination of radiation and heat. The limited effect of the temperature alone was confirmed by the fact that water from dark-bottles reactors retained its infectivity capacity.

On the other hand, it was reported that, although certain modifications in viral proteins can occur, the exposure to UV irradiation and ROS damages by $^1O_2$ seems to be a more strong-genome damage component, since it transforms RNA itself by dimers or RNA–RNA and RNA–protein cross-links (Wigginton et al., 2012). This suggests that RT-qPCR amplification loss by UV genome damages could be an appropriate proxy for SODIS evaluations in certain conditions, like low or middle temperatures, when damages to viral capsids are minimal.

FENTON AND PHOTO-FENTON

Principles of the Methods

Advanced oxidation processes (AOPs) have been recently considered as promising treatments for water disinfection (Malato et al., 2009). Although AOPs can use different reacting systems, all of them are based on the physicochemical generation of powerful *in situ* transitory species, principally hydroxyl radicals ($^{\bullet}OH$), highly efficient in organic matter oxidation. Among different AOPs, homogeneous Fenton and photo-Fenton processes have attracted great interest in the field of solar water disinfection (Malato et al., 2009).

Homogeneous Fenton process is based on the formation of $^{\bullet}OH$ when, in absence of light, hydrogen peroxide (H_2O_2) is decomposed by Fe^{2+} ions present in the aqueous (aq) phase (Chong et al., 2010; Pignatello et al., 2006):

$$Fe^{2+} + H_2O_2 \rightarrow Fe^{3+} + OH^- + {}^{\bullet}OH$$

$$Fe^{3+} + H_2O_2 \rightarrow Fe^{2+} + HO_2^{\bullet} + H^+$$

$$Fe^{3+} + HO_2^{\bullet} \rightarrow Fe^{2+} + O_2 + H^+$$

In presence of UV irradiation wavelengths of up to 600 nm, the reaction is improved compared with dark conditions by the additional production of $^{\bullet}OH$ (Mailhot et al., 2002):

$$Fe(OH)^{2+} + h\nu \rightarrow Fe^{2+} + {}^{\bullet}OH$$

This occurs by the efficient photo-reduction of the main dissolved Fe^{3+} species, the aquacomplex of Fe^{3+}, mainly $Fe(OH)^{2+}$, to Fe^{2+} that close the photo-catalytic cycle, reset the Fenton reaction and lead to continuous $^{\bullet}OH$ generation while eliminating the need for further addition of iron (Pignatello et al., 2006). The photo-Fenton process is affected by the pH, the amount of dissolved iron, H_2O_2 concentration, temperature, irradiance and the presence of other ions (Chong et al., 2010; Malato et al., 2009).

The ability of homogeneous photo-Fenton to remove certain microorganisms, including bacteria, fungi and protozoa, in water and wastewater has been demonstrated (Argulló-Barceló et al., 2013; García-Fernández et al., 2012; Ortega-Gómez et al., 2012; Polo-López et al., 2012; Spuhler et al., 2010). However, studies evaluating the efficacy of this process against virus remain scarce and restricted to some indicator viruses, like bacteriophages, and using UV lamps and simulated sunlight (Nieto-Juarez et al., 2010), and also using concentrated solar light (Argulló-Barceló et al., 2013).

Water Disinfection Experiments

Fenton and solar photo-Fenton experiences were carried out at Plataforma Solar de Almería, Tabernas dessert (Spain) to evaluate the effectiveness of these methods for the disinfection of MNV (as human NoV surrogate) in water under field conditions.

Fenton and photo-Fenton were performed in 250-mL DURAN-glass (Schott) batch stirred bottle reactors filled with 200 ml of sterile distilled water, contaminated with MNV at an initial concentration of 10^5 pfu/ml. For photo-Fenton experiments the bottle reactors were spiked with three

Fe^{2+} concentrations (2.5, 5 and 10 mg/L) using ferrous sulphate heptahydrate ($FeSO_4 \cdot 7H_2O$, PANREAC, Spain) and three H_2O_2 concentrations (5, 10 and 20 mg/l) at 35 wt %. pH was not adjusted in photo-Fenton experiments; it ranged from 3.6 to 6.3 depending on the different amounts of added ferrous sulphate for each experimental condition (Table 2). In parallel, Fenton reactors wrapped in aluminium foil (to maintain the reactors in the darkness and discriminate the sunlight radiation), and control reactors (bottles contaminated with MNV and stored in the dark) were included in the experiments. The Fenton and photo-Fenton reactors were exposed to sunlight for 6h. Disinfection kinetics were determined by RT-real time PCR (RT-qPCR) and infectivity assays as described by Polo et al., (2015).

Maximum local noon UV irradiances recorded for MNV experiments within the solar exposure period were 17.81 W/m^2 (Figure 2). The accumulated UV dose and Q_{UV} at the end of the exposure period were 314.5 kJ/m^2 and 14.94 kJ/l (Figure 3). Maximum global sunlight irradiance values for MNV experiments were 300.16 W×m^2 (average per hour 136.41 W × m^2).

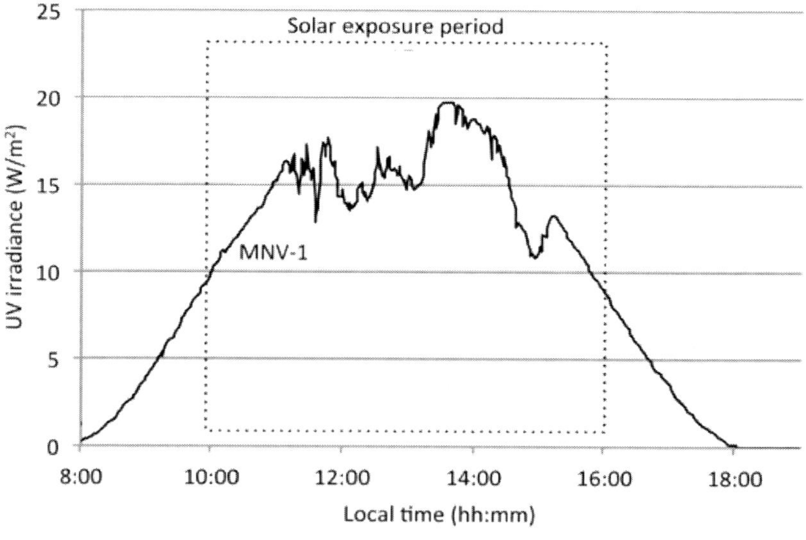

Figure 2. Average solar UV irradiance (295-385 nm) along the solar exposure period performed with MNV.

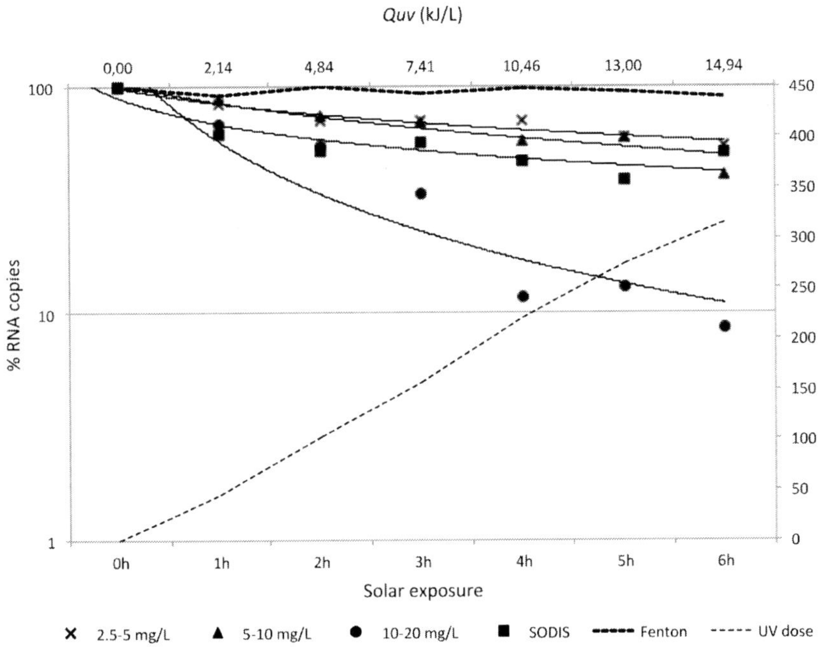

Figure 3. MNV RNA removal kinetics along the exposure period under the different treatments assayed: Fenton (with 10-20 mg/L of Fe^{2+}-H_2O_2), SODIS and solar photo-Fenton (with three different concentrations: 2.5-5; 5-10 and 10-20 mg/L Fe^{2+}-H_2O_2). Results are expressed as % of viral RNA copies/100 mL. Each data point is the average of three replicates. Solar UV dose (kJ/m^2) and cumulative UV radiation (Q_{UV}, kJ/l) are also shown in secondary axis. R^2 for each trend-line are: 0.80 (SODIS), 0.95 (2.5-5 mg/l), 0.93 (5-10 mg/l), 0.87 (10-20 mg/L) and 0.97 (Fenton).

Average water temperatures reached in bottles along the exposure time for MNV experiments, were 24.1°C (Fenton) and 26.2, 26.1 and 26.4°C for solar photo-Fenton (2.5-5, 5-10 and 10-20 mg/l Fe^{2+} and H_2O_2, respectively). The minimum and maximum water temperatures besides initial and final concentration of pH, dissolved oxygen, dissolved Fe and H_2O_2 recorded within the glass bottles in MNV experiments for each treatment are shown in Table 2. Average of viral reductions at the end of exposure time were 0.05 Log-units (10.1 %) in Fenton, and 0.27, 0.39 and 1.06 Log-units (45.3, 60.3 and 91.4 %) in photo-Fenton for 2.5-5, 5-10 and 10-20 mg/l Fe^{2+} and H_2O_2 (Figure 3; Table 3).

Table 2. Minimum (min) and maximum (max) temperatures (ºC) and initial (i) and final (f) values of pH, dissolved oxygen (O_2), total dissolved Fe, H_2O_2 reached for each treatment assayed (SODIS, photo-Fenton and Fenton) along the study period in MNV trials

Treatment Fe^{+2}/H_2O_2 (mg/L)	T (ºC) *min*/*max*	pH*i*/pH*f*	O_2i/O_2f (mg/L)	Fe*i*/Fe*f* (mg/L)	H_2O_2i/H_2O_2f (mg/L)
SODIS (0/0)	21.2/27.0	6.1/6.2	7.3/7.0	–/–	–/–
Pho-F (2.5/5)	21.2/28.9	4.1/6.3	7.6/6.8	2.71/0.63	5.15/4.11
Pho-F (5/10)	21.2/28.5	3.9/5.4	7.8/6.7	4.04/1.89	6.70/3.60
Pho-F (10/20)	21.2/29.6	3.7/4.1	8.7/6.7	7.38/6.57	11.36/3.60
Fenton (10/20)	21.2/25.8	3.7/3.8	9.2/5.4	8.94/7.38	10.84/3.60

The lower efficacy of RNA removal after the first hours indicates that an accumulation of damages is required to clear injury the virus. At first hours, photo-Fenton reagents produce the main damage, probably because H_2O_2 is mostly consumed in the period of the first 2-3h. In addition, Fe^{2+} could not be transformed to $Fe^{3+}$and at the same time, the reduction of Fe^{3+} by light occurs until all iron is transformed into Fe2+, which progressively stops the photo-Fenton reaction.

Regarding the infectivity assays, samples under photo-Fenton 5-10 mg/l Fe^{2+} -H_2O_2 conditions showed a lack of CPE after 5 and 6h with a 20% of infectivity decrease. Also, 20% of infectivity decrease after 3h and 40% after 4, 5 and 6h of solar exposure were observed in samples with 10-20 mg/l Fe^{2+}-H_2O_2 under photo-Fenton conditions. Meanwhile, samples from bottle reactors with lowest concentration of Fe^{2+}-H_2O_2 (2.5-5 mg/l) maintained the infectivity capacity along the photo-Fenton experience.

This despicable reduction, once again, could be interpreted by the high stability in hard environments presented from this kind of enteric non-enveloped viruses. Differences in the inactivation rates are explained by the change of iron speciation in water at different pH. Iron is more stable and soluble at more acidic pH, that promotes the presence of $Fe(OH)^{2+}$ dominant species and enhance the generation of $^{\bullet}OH$ radicals.

Table 3. Initial and final MNV quantification obtained in each treatment assayed (SODIS, photo-Fenton and Fenton) expressed as RNAc/100ml (average of three replicates with the correspondent Standard deviation shown in brackets) and Log-units of viral removal at the end of the exposure period (6 h)

Treatment Fe^{+2}/H_2O_2 (mg/L)	RNAc/100mL *i*(0h)	RNAc/100mL *f*(6h)	Log units
SODIS (0/0)	$1.4(\pm0.7)\times10^7$	$7.2(\pm0.7)\times10^6$	0.29
Pho-F (2.5/5)	$1.8(\pm0.2)\times10^7$	$1.0(\pm0.2)\times10^7$	0.27
Pho-F (5/10)	$4.1(\pm1.7)\times10^7$	$1.7(\pm0.5)\times10^7$	0.39
Pho-F (10/20)	$2.8(\pm0.4)\times10^7$	$2.4(\pm1.6)\times10^6$	1.06
Fenton (10/20)	$1.8(\pm0.3)\times10^7$	$1.6(\pm0.8)\times10^7$	0.05

Discussion

The growth of population and industrialization increase the demand on local and regional water supplies required for domestic, irrigation and industrial uses. The lack of access to adequately treated drinking water and sanitation in developing countries continue to be the most important issue to prevent transmission of NoV. In developing countries people that live in a household is higher than those that live in industrialized countries and the water sanitation methods are less than optimal (Colombati et al., 2009; Cronin et al., 2009; Nkwan, 2009; Montgomery and Elimelech, 2007). Poor quality of water hygiene could increase the persistence and the circulation of enteric viruses (Ayukekbong et al., 2015) and, therefore, improvement of good hygiene measures like avoiding contaminated water and food could be one of the first steps to achieve this aim (Koopman and Duzier, 2004). In countries with major economic level and with better water sanitation and hygiene, waterborne diseases are still being documented. A high percentage of NoV infections in USA (~60%) is estimated to be due to contaminated water (Mead et al., 1999). A study conducted between the years 1971 and 2006, also in USA, established that

87.2% of diseases associated with drinking water had their origin in public water supplies (Craun et al., 2010), many of which occurring as result of untreated or inappropriate treated water consumption or deficiencies in distribution systems.

Use of wastewater, sewage sludge and fecal material, diluted or raw, is a widespread practice in agriculture as a source of nutrients for crops (Jiménez et al., 2010). This practice plays a very important role in the worldwide transmission of enteric pathogens. The increase in the number of outbreaks associated with fruits and vegetables during the last two decades corresponds to an increase in their demand, expansion in a globalized market, and the way of consumption, raw or minimally processed (Cheong et al., 2009; Gerba and Choi, 2006; Sivapalasingam et al., 2004). A recent example is the case of a NoV outbreak occurred in Germany associated to strawberries imported from China that were (supposedly) cultivated with untreated wastewater (Mäde et al., 2013).

Energetic and economic costs associated to disinfection techniques are the principal disadvantage for the implementation of such procedures in developing countries. Therefore, alternative methods for water disinfection are required. HWTS represent low cost- low energy consumption, environmentally sustainable alternatives, with no adverse health effects. Evidences mentioned in this chapter offer an objective vision about SODIS, Fenton and photo-Fenton effectiveness for water disinfection in samples by simple exposure to natural sunlight conditions.

SODIS assays achieved a significant reduction of MNV after 8h of solar exposition, together with an infectivity decrease after 4h. In addition, comparisons between different conditions of reactor bottles determined sun radiation over temperature as principal factor for viral reduction. Despite that the average disinfection seems relatively low, this HWTS should not be discarded as alternative for the reduction of enteric virus transmission. In fact, increasing temperature would raise the efficacy of the process. This can be an important factor to be considered due to the fact that in populated areas with high incidence of enteric virus transmission, optimal temperatures and irradiation to improve SODIS methods could be easily reached. Also the efficiency of this SODIS basic protocol could be improved by additional developments, as placing reactor bottles on reflective surfaces (Mani et al., 2006; Kehoe et al., 2001), painting the underside of SODIS bottle reactors black (Sommer et al., 1997), increase the initial dissolved oxygen levels by shaking the bottles (Reed, 2004; Reed et al., 1997), filtering the water sample before filling the bottle reactors or using solar collectors, not only for enhancing solar disinfection but also for treatment of higher water volumes (Fontán-Sainz et al., 2012).

On the other hand, AOPs should be considered as useful, sustainable and viable alternatives to guarantee acceptable water sources, in terms of microbiological quality, to other purposes as food irrigation. Fenton and photo-Fenton are interesting methods for this kind of accomplishment. As indicated by Ortega-Gómez et al., (2015), the mechanism of virus inactivation of these techniques can be explained as a cascade of multiple steps including i) physical adsorption of the metal oxides onto the virus surface through electrostatic interactions; ii) solar radiation causes direct-endogenous damage in virus proteins and nucleic acids which affect the binding of the virus to the host cells; iii) the Fe^{3+}-virus complex acts as a photosensitizer, favouring solar absorption and generating Fe^{2+} and $^{\bullet}OH$ radicals, which give rise to indirect-endogenous virus damage; and iv) the exposure to solar light of the organic matter present in natural waters, forms a triplet excited state of dissolved organic matter that in presence of oxygen generates ROS, leadin to virus inactivation (Ruales-Lonfat et al., 2014; Nieto-Juarez and Kohn, 2013).

Results presented above show that under natural solar conditions and by using Fe^{2+} and H_2O_2, viral quantifications can be significantly reduced. Best disinfection results were achieved with 10 mg/l and 20 mg/l of Fe^{2+} and H_2O_2 concentrations together with a decrease of 40% in infectivity. It is interesting to highlight that only suboptimal photo-Fenton conditions could be acomplished in these experiences and, therefore, higher water disinfection can be expected in areas where natural conditions are more favourable (higher sunlight radiation and temperatures).

Solar photo-Fenton water disinfection for drinking or recreational purposes could be considerably improved using higher values of radiation and water temperatures, while reducing the concentration of reagents.

Therefore, future research focused on better concentrations of photo-Fenton reagents is needed to optimize the effectivity of the process bearing in mind that, due to the toxicity of H_2O_2 or Fe^{2+}, water treated by certain photo-Fenton concentration would be not adequate for human consumption. For the World Health Organization there is not limit value in iron ingestion, nonetheless it is established a provisional level at 2 mg/l for drinking water (WHO, 2011). Europe normative is more restrictive with a 200 µg/l (Anon, 1998). Nevertheless guidelines of safety standards for wastewater reuse are different depending on the intended final use, being more tolerant for irrigation water.

CONCLUSION

The overexploitation of fresh water resources and the increase in human population are leading to one of the most serious problems facing the world: water scarcity and the lack of access to safe drinking water. The most important issues related to water resources are the disinfection of drinking water as well as wastewater disinfection for multiple reuses, including agriculture or aquifer refilling, among others.

Ensuring the safety and quality of drinking water is essential in prevention of waterborne diseases. All the methods mentioned in this chapter could be employed for the reduction of NoV transmission in developing areas, where enteric virus infections through consumption of untreated or poorly treated water is still highly prevalent.

On the other hand, the absence of technical and economic resources in developing countries makes the reuse of raw or diluted wastewater a long-term future strategy. In this sense, photo-Fenton processes could be a sustainable and low cost solution to avoid, at least in part, health impacts from wastewater irrigation and other treated wastewater re-uses. One of the major challenges for scientists using photo-Fenton as a water treatment for drinking or recreational waters is to find the conditions of greatest process efficiency at neutral or near-neutral pH.

Further research in other types of water or wastewater is needed to evaluate the efficacy and limitations of these processes from a virological point of view. Nevertheless, results presented here suggest SODIS and solar photo-Fenton technologies as feasible alternatives for water treatment.

REFERENCES

Agulló-Barceló M, Polo-López MI, Lucena F, Jofre J, Fernández-Ibáñez P. 2013. Solar advanced oxidation processes as disinfection tertiary treatments for real wastewater: implications for water reclamation. *Appl Cat B: Environ* 136: 341-350.

Atmar RL, Estes MK, 2001. Diagnosis of noncultivatable gastroenteritis viruses, the human caliciviruses. *Clin Microbiol Rev*, 14, 15-37.

Atmar RL, Opekun AR, Gilger MA, Estes MK, Crawford SE, Neill FH, Ramani S, Hill H, Ferreira J, Graham DY. 2014. Determination of the 50% human infectious dose for norwalk virus. *J Infect Dis* 209: 1016–1022.

Ayukekbong JA, Mesumbe HN, Oyero OG, Lindh M, Bergström, T. 2015. Role of noroviruses as aetiological agents of diarrhoea in developing countries. *J Gen Virol* 96: 1983-1999.

Bae J, Schwab KJ. 2008. Evaluation of murine norovirus, feline calicivirus, poliovirus, and MS2 as surrogates for human norovirus in a model of viral persistence in surface water and groundwater. *Appl Environ Microbiol* 74: 477-484.

Baehner F, Bogaerts H, Goodwin R. 2016. Vaccines against norovirus: state of the art trials in children and adults. *Clin Microbiol Infect* 22: S136-S139.

Baert L, Wobus CE, Van Coillie E, Thackray LB, Debevere J, Uyttendaele M. 2008. Detection of murine norovirus 1 by using plaque assay, transfection assay, and real-time reverse transcription-PCR before and after heat exposure. *Appl Environ Microbiol* 74:543-546.

Barreira DM, Ferreira MS, Fumian TM, Checon R, de Sadovsky AD, Leite JP, Miagostovich MP, Spano LC. 2010. Viral load and genotypes of noroviruses in symptomatic and asymptomatic children in Southeastern Brazil. *J Clin Virol* 47: 60-64.

Bosch A. 2010. Human enteric viruses in the water environment: a minireview. *Int Microbiol*, 1: 191-196.

Boschi-Pinto C, Velebit L, Shibuya K. 2008. Estimating child mortality due to diarrhoea in developing countries. *Bull World Health Organ* 86: 710-717.

Boyle M, Sichel C, Fernández-Ibáñez P, Arias-Quiroz GB, Iriarte-Puna M, Mercado A, Ubomba-Jaswa E, McGuigan KG. 2008. Bactericidal effect of solar water disinfection under real sunlight conditions. *Appl Environ Microbiol* 74: 2997-3001.

Chan MC, Sung JJ, Lam RK, Chan PK, Lee NL, Lai RW, Leung WK. 2006. Fecal Viral Load and Norovirus-associated Gastroenteritis. *Emerg Infect Dis* 12: 1278-1280.

Cheong S, Lee C, Song SW, Choi WC, Lee CH, Kim SJ. 2009. Enteric viruses in raw vegetables and groundwater used for irrigation in South Korea. *Appl Environ Microbiol* 75: 7745-7751.

Chong MN, Jin B, Chow CW, Saint C. 2010. Recent developments in photocatalytic water treatment technology: a review. *Water res* 44: 2997-3027.

Colombatti R, Vieira CS, Bassani F, Cristofoli R, Coin A, Bertinato L, Riccardi F. 2009. Contamination of drinking water sources during the rainy season in an urban post-conflict community in Guinea Bissau: implications for sanitation priority. *Afr J Med Med Sci* 38: 155-161.

Craun GF, Brunkard JM, Yoder JS, Roberts VA, Carpenter J, Wade T, Calderon RL, Roberts, JM, Beach MJ, Roy SL. 2010. Causes of outbreaks associated with drinking water in the United States from 1971 to 2006. *Clin Microbiol Rev* 23: 507-528.

Cronin AA, Shrestha D, Spiegel P, Gore F, Hering H. 2009. Quantifying the burden of disease associated with inadequate provision of water and sanitation in selected sub-Saharan refugee camps. *J Water Health* 7: 557-568.

Eisenstark A. 1987. Mutagenic and lethal effects of near-ultraviolet radiation (290–400 nm) on bacteria and phage. *Environ Mol Mutagen* 10: 317-337.

Ettayebi K, Crawford SE, Murakami K, Broughman JR, Karandikar U, Tenge VR, Neill FH, Blutt SE, Zeng XL, Qu L, Kou B, Opekun AR, Burrin D, Graham DY, Ramani S, Atmar RL, Estes MK. 2016. Replication of human noroviruses in stem cell–derived human enteroids. *Science,* 353: 1387-1393.

Fewtrell L, Bartram J, 2001. Water Quality: Guidelines, Standards & Health. IWA publishing.

Fontán-Sainz M, Gómez-Couso H, Fernández-Ibáñez P, Ares-Mazás E. 2012. Evaluation of the solar water disinfection process (SODIS) against Cryptosporidium parvum using a 25-L static solar reactor fitted with a compound parabolic collector (CPC). *Am J Trop Med Hyg* 86: 223-228.

Gall AM, Mariñas BJ, Lu Y, Shisler JL. 2015. Waterborne viruses: a barrier to safe drinking water. *PLoS Pathog* 11: e1004867.

García-Fernández I, Polo-López MI, Oller I, Fernández-Ibáñez, P. 2012. Bacteria and fungi inactivation using Fe 3+/sunlight, H 2O2/sunlight and near neutral photo-Fenton: A comparative study. *Appl Cat B: Environ* 121: 20-29.

Gerba CP, Choi CY. 2006. Role of irrigation water in crop contamination by viruses. In Viruses in foods: Springer US.

Goodsell DS. 2001. The molecular perspective: ultraviolet light and pyrimidine dimers. *Oncologist* 6: 298-299.

Haas CN, Rose JB, Gerba C, Regli S. 1993. Risk assessment of virus in drinking water. *Risk Anal* 13: 545-552.

Head MG, Lopman BA. 2016. Norovirus in 2016 – Emesis aplenty but clear signs of progress. *J Infect Dis* 213: S1-S2.

Hewitt J, Rivera-Aban M, Greening GE. 2009. Evaluation of murine norovirus as a surrogate for human norovirus and hepatitis A virus in heat inactivation studies. *J Appl Microbiol* 107: 65-71.

Jagger J. 1981. Near-UV radiation effects on microorganisms. *Photochem Photobiol* 34: 761-768.

Jiménez B, Drechsel P, Koné D, Bahri A, Raschid-Sally L, Qadir M. 2010. Wastewater, sludge and excreta use in developing countries: an overview. In : Dreschsel P, Scott CA, Rashid-Sally L, Redwood M, Bahri A (Eds.) Wastewater Irrigation and Health. Assessing and Mitigating Risk in Low-income Countries. Earthscan; London, UK.

Kapikian AZ, Wyatt RG, Dolin R, Thornhill TS, Kalica AR, Chanock RM. 1972. Visualization by immune electron microscopy of a 27-nm particle associated with acute infectious nonbacterial gastroenteritis. *J Virol* 10: 1075-1081.

Kehoe SC, Joyce TM, Ibrahim P, Gillespie JB, Shahar, RA McGuigan KG. 2001. Effect of agitation, turbidity, aluminium foil reflectors and container volume on the inactivation efficiency of batch-process solar disinfectors. *Water res* 35: 1061-1065.

Kocher J, Yuan L. 2015. Norovirus vaccines and potential antinorovirus drugs: recent advances and future perspectives. *Future Virol* 10: 899-913.

Koopmans M, Duizer E, 2004. Foodborne viruses: an emerging problem. *Int J Food Microbiol* 90: 23-41.

Lopman BA, Grassly NC. 2016. Pediatric Norovirus in Developing Countries: A Picture Slowly Comes Into Focus. *Clin Infect Dis*, ciw078.

Mäde D, Trübner K, Neubert E, Höhne M, Johne R. 2013. Detection and typing of norovirus from frozen strawberries involved in a large-scale gastroenteritis outbreak in Germany, *Food Environ Virol* 5: 162-168.

Mailhot G, Sarakha M, Lavedrine B, Caceres J, Malato S. 2002. Fe (III)-solar light induced degradation of diethyl phthalate (DEP) in aqueous solutions. *Chemosphere,* 49: 525-532.

Malato S, Fernández-Ibáñez P, Maldonado MI, Blanco J, Gernjak W. 2009. Decontamination and disinfection of water by solar photocatalysis: recent overview and trends. *Catal Today* 147: 1-59.

Mani SK, Kanjur R, Singh ISB, Reed RH. 2006. Comparative effectiveness of solar disinfection using small-scale batch reactors with reflective, absorptive and transmissive rear surfaces. *Water res* 40: 721-727.

Maunula L, Söderberg K, Vahtera H, Vuorilehto VP, von Bonsdorff CH, Valtari M, Laakso T, Lahti K, 2012. Presence of human noro-and adenoviruses in river and treated wastewater, a longitudinal study and method comparison. *J Water Health*, 10, 87-99.

McGuigan KG, Conroy RM, Mosler HJ, du Preez M, Ubomba-Jaswa E, Fernandez-Ibanez P. 2012. Solar water disinfection (SODIS): a review from bench-top to roof-top. *J Hazard Mater* 235: 29-46.

McGuigan KG, Joyce TM, Conroy RM, Gillespie JB, Elmore-Meegan M. 1998. Solar disinfection of drinking water contained in transparent plastic bottles: characterizing the bacterial inactivation process. *J Appl Microbiol* 84: 1138-1148.

Mead PS, Slutsker L, Dietz V, McCaig LF, Bresee JS, Shapiro C, Griffin PM, Tauxe RV. 1999. Food-related illness and death in the United States. *Emerg Infect Dis* 5: 607.

Montgomery MA, Elimelech M. 2007. Water and sanitation in developing countries: including health in the equation. *Environ Sci Technol* 41: 17-24.

Nieto-Juarez JI, Kohn T. 2013. Virus removal and inactivation by iron (hydr) oxide-mediated Fenton-like processes under sunlight and in the dark. *Photochem Photobiol Sci* 12: 1596-1605.

Nieto-Juarez JI, Pierzchła K, Sienkiewicz A, Kohn T. 2010. Inactivation of MS2 coliphage in Fenton and Fenton-like systems: role of transition metals, hydrogen peroxide and sunlight. *Environ Sci Technol* 44: 3351-3356.

Nkwan J. 2009. Buea Town: water, sanitation, millennium development goals, and local realities in Cameroon. *Can J Infect Control* 24: 28-29.

Nuanualsuwan S, Cliver DO. 2003. Capsid functions of inactivated human picornaviruses and feline calicivirus. *Appl Environ Microbiol* 69: 350-357.

Ortega-Gómez E, Fernández-Ibáñez P, Martín MB, Polo-López MI, García BE, Pérez JS. 2012. Water disinfection using photo-Fenton: Effect of temperature on Enterococcus faecalis survival. *Water res* 46: 6154-6162.

Ortega-Gómez E, Martín MB, Carratalà A, Ibañez PF, Pérez JS, Pulgarín C. 2015. Principal parameters affecting virus inactivation by the solar photo-Fenton process at neutral pH and μM concentrations of H 2 O 2 and Fe 2+/3+. *App Catal B: Environ* 174: 395-402.

Patel MM, Widdowson MA, Glass RI, Akazawa K, Vinjé J, Parashar UD. 2008. Systematic literature review of role of noroviruses in sporadic gastroenteritis. *Emerg Infect Dis* 14: 1224–121.

Pignatello JJ, Oliveros E, MacKay A. 2006. Advanced oxidation processes for organic contaminant destruction based on the Fenton reaction and related chemistry. *Crit Rev Environ Sci Technol* 36: 1-84.

Polo D, García-Fernández I, Fernández-Ibáñez P, Romalde JL. 2015. Solar water disinfection (SODIS): Impact on hepatitis A virus and on a human Norovirus surrogate under natural solar conditions. Int *Microbiol* 18: 41-49.

Polo-López MI, García-Fernández I, Velegraki T, Katsoni A, Oller I, Mantzavinos D, Fernández-Ibáñez P. 2012. Mild solar photo-Fenton: An effective tool for the removal of Fusarium from simulated municipal effluents. *Appl Catal B: Environ*, 111: 545-554.

Pringle K, Lopman B, Vega E, Vinje J, Parashar UD, Hall AJ. 2015. Noroviruses: epidemiology, immunity and prospects for prevention. *Future Microbiol* 10: 53-67.

Prüss-Üstün A, Bos R, Gore F, Bartram J. 2008. Safer water, better health: costs, benefits and sustainability of interventions to protect and promote health. WHO Press. World Health Organization, Geneva, Switzerland.

Reed RH. 1997. Solar inactivation of faecal bacteria in water: the critical role of oxygen. *Lett Appl Microbiol* 24: 276-280.

Reed RH. 2004. The inactivation of microbes by sunlight: solar disinfection as a water treatment process. *Adv Appl Microbiol* 54: 333-365.

Redd RH, Mani SK, Meyer V. 2000. Solar phto-oxidative disinfection of drinking water: preliminary field observations. *Lett Appl Microbiol* 30: 432-436.

Rodriguez-Lazaro D, Cook N, Ruggeri FM, Sellwood J, Nasser A, Nascimento MSJ, D'Agostino M, Santos R, Saiz JC, Rzezutka A, Bosch A, Girones R, Carducci A, Muscillo M, Kovac K, Diez-Valcarce M, Vantarakis A, von Bonsdorff CH, de Roda Husman AM, Hernandez M, van der Poel WHM, A. 2012. Virus hazards from food, water and other contaminated environments. *FEMS Microbiol Rev* 36: 786-814.

Ruales-Lonfat C, Benítez N, Sienkiewicz A, Pulgarín C. 2014. Deleterious effect of homogeneous and heterogeneous near-neutral photo-Fenton system on Escherichia coli. Comparison with photo-catalytic action of TiO_2 during cell envelope disruption. *App Catal B: Environ* 160: 286-297.

Scallan E, Hoekstra RM, Angulo FJ, Tauxe RV, Widdowson MA, Roy SL, Jones JL, Griffin PM. 2011. Foodborne illness acquired in the United States – major pathogens. *Emerg Infect Dis* 17: 7–15.

Sivapalasingam S, Friedman CR, Cohen L, Tauxe RV. 2004. Fresh produce: a growing cause of outbreaks of foodborne illness in the United States, 1973 through 1997. *J Food Prot* 67: 2342-2353.

Sommer B, Marino A, Solarte M, Dierolf C, Valiente C, Mora D, Rechsteiner R, Setters P, Wirojanaguds W, Ajarmeh H, Al-Hassan A, Wegelin M. 1997. SODIS- an emerging water treatment process. *J Water SRT – Aqua* 46: 127-137.

Spuhler D, Rengifo-Herrera JA, Pulgarin C. 2010. The effect of Fe 2+, Fe 3+, H 2 O 2 and the photo-Fenton reagent at near neutral pH on the solar disinfection (SODIS) at low temperatures of water containing Escherichia coli K12. *App Catal B: Environ* 96: 126-141.

Teunis PF, Moe CL, Liu P, Miller SE, Lindesmith L, Baric RS, Le Pendu J, Calderon RL. 2008. Norwalk virus: how infectious is it?. *Journal of medical virology*, 80, 1468-1476.

UNICEF. 2012. Pneumonia and diarrhoea: Tackling the deadliest diseases for the world's poorest children. New York, NY: Statistics and Monitoring Section- Division of Policy and Strategy.

Vaughan G, Rossi LMG, Forbi JC, de Paula VS, Purdy MA, Xia G, Khudyakov YE. 2014. Hepatitis A virus: host interactions, molecular epidemiology and evolution. *Infect Genet Evol* 21: 227-243.

Vinjé J. 2015. Advances in laboratory methods for detection and typing of norovirus. *J Clin Microbiol* 53: 373-381.

Wegelin M, Canonica S, Mechsner K, Fleischmann T, Pesaro F, Metzler A. 1994. Solar water disinfection: scope of the process and analysis of radiation experiments. *J Water SRT – Aqua* 43: 154-169.

WHO 2011. Guidelines for Drinking-water Quality. WHO Press. World Health Organization, Geneva, Switzerland.

WHO, UNICEF. 2014. Progress on drinking water and sanitation: 2014 Update. WHO Press. World Health Organization, Geneva, Switzerland.

Wigginton KR, Pecson BM, Sigstam T, Bosshard F, Khon T. 2012. Virus inactivation mechanisms: impact of disinfectants on virus function and structural integrity. *Environ Sci Technol* 46: 12069-12078

Zheng DP, Ando T, Fankhauser RL, Beard RS, Glass RI, Monroe SS. 2006. Norovirus classification and proposed strain nomenclature. *Virology* 346: 312-323.

In: Noroviruses
Editor: Jesús L. Romalde ISBN: 978-1-53611-939-8

Chapter 7

HIGH PRESSURE PROCESSING'S POTENTIAL TO INACTIVATE NOROVIRUSES AND OTHER FOODBORNE VIRUSES

David H. Kingsley*
U.S. Department of Agriculture, Agricultural Research Service,
Food Safety and Interventions Technology Research Unit,
W. W. Baker Center, Delaware State University, Dover, DE, US

ABSTRACT

High pressure processing (HPP) can inactivate human norovirus. However, all viruses are not equally susceptible to HPP. Pressure treatment parameters such as required pressure levels, initial pressurization temperatures, and pressurization times substantially affect inactivation. How food matrix characteristics affect different viruses is discussed and the specific influence of water content, salt, sugar, and pH are described. The mechanism by which HPP inactivates viruses and other pathogens is discussed. While specific pathogen inactivation within

* Corresponding Author address: U.S. Department of Agriculture, Agricultural Research Service, Food Safety and Interventions Technology Research Unit, W. W. Baker Center, Delaware State University, Dover, DE 19901, Email: David.Kingsley@ARS.USDA.GOV.

a specific food will require detailed studies to characterize and validate inactivation, this chapter provides a foundation for making basic predictions for norovirus virus inactivation by HPP.

Keywords: high pressure processing, norovirus, hepatitis A virus, hepatitis E virus, shellfish, vegetables

INTRODUCTION

Human noroviruses (HuNoVs) are the most frequent cause of foodborne illness in the developed world (Koo et al., 2010) and, in the U.S., viruses may cause as much as 59% of foodborne illness (Scallan et al., 2011). These viruses are very infectious and typically shed in large quantities. While HuNoVs are by far the most prevalent from an illness incidence perspective (Scallan et al., 2011), there are a plethora of human viruses that can be transmitted by the fecal-oral route, and therefore, can be transmitted as a result of food contamination. Among these are hepatitis A virus (HAV), hepatitis E virus (HEV), rotavirus, astrovirus, adenovirus, sapoviruses and enteroviruses. Most foodborne viruses are transmitted via the fecal-oral route with the original source being human feces. However, HuNoV can also be spread by aerosolized vomitus contamination (Makison-Booth, 2014; Wu et al., 2005). Foods commonly eaten uncooked such as shellfish, which concentrate waterborne viruses, and fresh produce can be considered higher risk foods for norovirus transmission (Amon et al., 2005; Baert et al., 2011; Bouwknegt et al., 2015). Fruits and vegetables can be contaminated at the preharvest stage through contact with contaminated irrigation water or fecal fertilizer in the field and during harvest, processing, and at retail as a result of bare hand contact (Carter, 2005; Loutreul et al., 2014). For shellfish, human waste pollution of growing waters is the principal source of viral contamination (Kingsley, 2013, 2014b). Generally speaking, foodborne viruses are very stable in the environment, remaining viable for several months, making their control very challenging. This is primarily due to the fact that they are almost exclusively nonenveloped, having only a rugged proteinaceous capsid. Furthermore, foodborne viruses are often resistant to low pH and can resist disinfection and sanitizers better than most bacteria (Cannon et al., 2006; Kingsley et al., 2014c). As a result of these challenges, there is need for nonthermal technologies that can inactivate noroviruses and other foodborne viruses.

High pressure processing (HPP) can be effective against viruses and is now becoming an increasingly common nonthermal food processing technology. In many cases this technology allows inactivation of pathogenic and spoilage microorganisms in foods with limited changes in flavor, texture, or color (Balasubramaniam et al., 2015; Kingsley, 2013). The range of commercial products utilizing HPP includes several fruit- and vegetable-based refrigerated food products such fruit juices, smoothies, jams and jellies, guacamole, tomato-based salsas, ready-to-eat meat products, as well as shellfish on-the-half-shell. (Balasubramaniam et al., 2015; Wang et al., 2016). New HPP-processed food products with enhanced safety and nutritional qualities now regularly appear in the marketplace. In the US, a couple of prominent franchised sandwich shops now utilize HPP for their deli meats as an assurance of food safety.

The first investigation into the use of high pressure as a non-thermal means of inactivating bacteria was performed by Bert Hite in the 1890s, who showed HPP could pasteurize milk (Hite, 1899). The first attempts at evaluating whether high pressure could inactivate a virus was performed by Hite and coworkers using tobacco mosaic virus (TMV; Giddings et al., 1929). TMV was found to be extremely resistant to pressure. Fortunately, studies performed over the better part of the last two decades have revealed that HPP is capable of effectively inactivating many foodborne viruses, making it a promising and practical nonthermal technology. The efficacy of HPP is substantially influenced by processing parameters such as pressure level, initial pressurization temperature, holding time, and nonprocessing parameters that relate to the virus' characteristics and barotolerance, as well as food matrix characteristics. Therefore, balancing the optimal parameters for enhancing virus inactivation while ensuring sensory and nutritional quality retention in foods is critical.

CHARACTERISTICS OF VIRUSES

Viruses cannot replicate independently. Typical foodborne viruses have no cellular structure but do have a protein coat, termed a capsid, that encases one or more molecules of either RNA or DNA. The capsid functions to mediate virus attachment, penetration and eventual release of the genetic material within the host cell. Noroviruses are less than 40 nm in diameter and viruses in general have diameters ranging from 10 nm to 400 nm (Prescott et al., 2002) making them considerably smaller than bacteria. Visualization of most viruses requires an electron microscope.

A foodborne virus is typically thought of as any virus that may be transmitted via the fecal-oral route and thus can be disseminated by contaminated food, or even water, to cause illness. Viruses of current importance are noroviruses, picornaviruses (hepatitis A virus, aichivirus, enteroviruses), hepatitis E virus, rotavirus, adenovirus and astrovirus (Grove et al., 2006; Kingsley, 2013). With the exception of adenovirus, which has double-stranded DNA, and rotavirus, which has double-stranded RNA, the other viruses are single-stranded, positive-sense RNA viruses.

Viruses are not involved in food spoilage and do not increase from temperature abuse because they cannot replicate in food products. However, since only a few particles can initiate an infection, and since extreme quantities of viral particles are shed in the stool (up to 10^{11} pfu/g feces), these pathogens have a high potential for transmission (Rzezutka and Cook, 2004). In addition to being environmentally stable, most foodborne viruses are more resistant to drying, freezing, acidic pH, and chemical sanitizers than bacteria (Baert et al., 2009; Kingsley et al., 2014c). Viruses also readily tolerate temperatures up to 60°C or even higher when sequestered within food matrices. Therefore, procedures designed to mitigate bacterial infections during food processing and preservation may not be effective against viral pathogens.

Human Noroviruses and Genetically-Related Surrogates

HuNoV was originally known as the Norwalk virus, since it was first visualized by electron microscopy (EM) from patient stools after an outbreak in an elementary school in Norwalk, Ohio in 1968 (Kapikian, 2000). Subsequently, noroviruses were called small round structured viruses (SRSVs), or Norwalk-like viruses, until more recently when these strains were collectively named noroviruses and grouped by genogroup (GI, GII or GIV), clade (a number after the genotype) and strain type (normally a location of an outbreak). *Norovirus* is now classified as a genus of the *Caliciviridae* family, although the capsid does not possess the cup-shaped depressions that are characteristic of the group (Green et al., 2000). The single-stranded (positive polarity) RNA genome is approximately 7.5 kb in length and contains three open reading frames (ORFs). These ORFs encode a large polyprotein, that is cleaved to generate several proteins, most of which are nonstructural proteins involved in virus replication, and the structural proteins VP1 and VP2 (Jiang et al., 1993). The NoV capsid is 27-40 nm in diameter and made up of 180 copies of the VP1 major capsid protein, forming an icosahedron shape (Prasad et al., 1999). The VP2 minor capsid protein is probably an RNA binding protein but may also contribute to capsid stability (Lin et al., 2014)

HuNoVs infect people of all ages. However, toddlers and senior citizens may be especially susceptible to infection. HuNoV infection can result in vomiting, diarrhea, nausea, low-grade fever, and abdominal cramping (Atmar and Estes, 2006). Cases typically become symptomatic 24–48 h after infection, and the illness typically resolves itself after 48–72 h (Teunis et al., 2008). In ordinary persons, illness is self-limited and mild; usually not requiring medical treatment. Fecal-oral spread is the main mode of food-borne transmission. However, transmission through infectious vomit, both by mechanical transmission from environmental surfaces (i.e., through hand/mouth contact) and aerosolization, is now understood to contribute to the rapid and extensive spread of the virus in closed settings, such as hospitals, hotels, cruise ships, commercial kitchens and day-care centers (Widdowson et al., 2005). Primary cases in an outbreak often result from exposure to contaminated food or water, whereas further propagation of the outbreak can be due to person-to-person spread. Foods such as sandwiches and salads are also common vehicles for HuNoV transmission, since these foods require manual handling but not subsequent cooking (Widdowson et al., 2005).

HuNoV can result from consumption of raw or undercooked oysters, mussels, cockles, and clams. Discharge of human waste in oyster harvest areas as a result of floods, or overboard discharge from boaters is frequently identified as the source of oyster contamination. Also illegal harvesting of shellfish from unapproved and unsanitary locations is a substantial problem. HuNoVs have been shown to interact specifically with carbohydrate structures in the bivalve digestive diverticula (LeGuyader et al., 2012), and to persist within phagocytic hemocytes (Provost et al., 2011). This suggests a virus-specific means of bioconcentration from contaminated waters, as well as a means by which the viruses can remain sequestered within shellfish tissue for several weeks. It is well understood that interventions that work for some bacteria within shellfish, such as depuration, are not effective for mitigating norovirus (Ueki et al., 2007).

Widespread food contamination before distribution and consumption is likely to go undetected because assays for detecting HuNoVs in foods are limited and not widely employed. Genome amplification-type methods, such as reverse-transcription-PCR, are currently the only practical method to detect HuNoV strains in food, environmental, and clinical samples. While there has been some recent progress toward a practical propagation model for HuNoV, routine propagation for testing foods remains enigmatic. Most inactivation studies have focused on propagable, genetically-related surrogates such as feline calicivirus (FCV), murine norovirus (MNV), and Tulane virus (TV; Farkas et al., 2008), which is a primate norovirus. Other less commonly used surrogates include San Miguel sea lion virus (Burkhardt et al., 2002) and canine calicivirus (Urbanucci et al. 2009). Direct assessment of HuNoV infectivity is currently only feasible using expensive and logistically-complicated human volunteer trials. Immunologically-naïve gnotobiotic piglets, delivered by caesarian section, are also being utilized as animal models for HuNoV (Cheetham et al., 2006: Lou et al., 2015). However, these animals are also expensive.

Other Viruses

Although much less common than norovirus, HAV is usually ranked second behind HuNoV on the list of most significant foodborne viruses, principally due to its medical seriousness. HAV is now uncommon in the developed world (Jacobsen and Koopman, 2004), but HAV can cause a serious illness that can be life-threatening. The primary symptom of hepatitis A illness is jaundice due to liver damage, which typically occurs 1-2 months after ingestion of virus-contaminated food or water (Fitzsimons et al., 2010). The virus is excreted in feces after being released from liver cells into the bile duct. For previously unexposed and unvaccinated persons over the age of 50, hepatitis A virus is associated with a mortality rate of approximately 1% (Fitzsimons et al., 2010). A highly effective vaccine has substantially reduced the prevalence of HAV in the US and the developed world (Scallan et al., 2011) over the past 20 years. However HAV remains prevalent in the developing world. For developing countries with high endemic HAV levels, this virus is not a major health priority since the population is largely immune and HAV, when acquired at a young age, is often without serious medical consequence. From the standpoint of global food trade, this creates a dangerous dichotomy where developing countries are largely immune to HAV, while previously unexposed populations in developed countries are highly susceptible to HAV infection. Recent examples of global trade products that have been implicated in HAV outbreaks include sun-dried tomatoes, pomegranate seeds, and imported scallops (Carvalho et al., 2012; Collier et al., 2014; Donnan et al., 2012).

Hepatitis A virus (HAV) was a member of the genus *Enterovirus*, but it is now placed in the *Hepatovirus* genus, within the family *Picornaviridae*. HAV is a small non-enveloped, single-stranded, positive-sense RNA virus with an estimated infectious dose of approximately 100 viral particles with a protein capsid of approximately 28-30 nm in diameter (Cliver and Matsui, 2002). Wild-type HAV is very difficult to propagate but fortunately a laboratory-adapted HAV strain HM-175 can be propagated *in vitro* (Cromeans et al., 1987).

Like other foodborne viruses, HAV is transmitted via the fecal-oral route, either by ingestion of contaminated food, water, or by direct contact with an infected person. The largest reported HAV outbreak occurred in 1988 in Shanghai, China, which was attributed to the consumption of contaminated raw clams that killed 32 people and resulted in nearly 300,000 HAV illness (Halliday et al., 1991). For domestically-harvested US shellfish, there have only been two outbreaks of HAV illness in the past 25 years. In both cases, illegally harvesting from unapproved growing waters was implicated (Desenclos et al., 1991; Shieh et al., 2007). In 2003, one of the more notable U.S. HAV outbreaks occurred in the Eastern U.S., causing more than 500 persons to contract the illness (Anon, 2003). Fecally-contaminated green onions from Latin America were confirmed as the source of the virus (Amon, et al., 2005; Wheeler et al., 2005).

While HuNoV and HAV are the main foodborne viruses of concern, there are a number of other fecal-oral viruses that can be transmitted through food. Among these are aichivirus, astroviruses, adenoviruses, hepatitis E virus, rotavirus, sapoviruses, and enteroviruses such as echoviruses, human parechoviruses (HPeVs) and coxsachieviruses (Fong and Lipp, 2005; Kingsley, 2016). Aichivirus (AiV) was first identified and isolated from a gastroenteritis patient in Aichi, Japan in 1989 (Yamshita et al., 1991; Yamashita et al., 2000). AiV outbreaks associated with shellfish have occurred in Europe and Asia (LeGuyader et al., 2008; Yamashita et al., 1998). Sapoviruses can also be transmitted by shellfish and are genetically-related to noroviruses but not as common (Iizuka et al., 2010; Ueki et al., 2010). Although a vaccine has reduced its incidence, rotavirus remains a concern among children <5 years old (Yi et al., 2016). The presence of rotaviruses has been reported in sewage and within bivalve shellfish (Cook et al., 2004; LeGuyader et al., 2008). Astroviruses have been identified as the cause of foodborne outbreaks and found within oysters implicated in outbreaks (Oishi et al., 1994; Yamashita et al., 1991). While respiratory illness is most commonly associated with adenovirus infection, serotypes 40 and 41 can cause gastroenteritis and are frequently found within human stool (Mena and Gerba, 2009). Ten percent of infantile gastroenteritis cases are thought to be caused by enteric adenoviruses, making this virus the second leading cause of diarrhea-related hospitalization in children (Fong and Lipp, 2005). Adenoviruses are also commonly isolated from shellfish (Pina et al., 1998).

Hepatitis E virus (HEV) is an emerging virus of some concern (Kupferschmidt, 2016). There are four HEV genotypes. Types 1 and 2 are transmitted person-to-person and causes a medically-serious illness with a 30-60 day incubation period and jaundice symptoms due to liver damage. Genotypes 3 and 4 are commonly associated with farm-raised and feral swine, game and other wild animals, and as a result, are considered zoonotic (Meng, 2011; Müller et al., 2016; Nantel-Fortier et al., 2016). It now appears that these strains can infect goats (DiMartino et al. 2016) and dairy herds (Huang et al., 2016). HEV is also waterborne, suggesting fruits and vegetables could easily be contaminated with this virus (Chen et al., 2016) and shellfish have been implicated in outbreaks (Gao et al., 2015; Grodzki et al., 2014; Mesquita et al., 2016). While zoonotic HEV is presently considered to be a mild infection for healthy persons, it can be a problem for people with underlying health issues (L'homme et al., 2016). Recent revelations indicate that, in some parts of Europe, as much as 20% of the population is seropositive for exposure to zoonotic hepatitis E virus (Kupferschmidt, 2016).

HPP and Foods

HPP can non-thermally inactivate foodborne pathogens, reduce total microbial load, and extend shelf life of food without affecting sensory quality (Balasubramaniam et al., 2015; Lopez-Caballero et al., 2000; He et al., 2002; Wang et al., 2016). Commercial applications of HPP include its use for beverages such as fruit juices, smoothies, jams and jellies as an alternative pasteurization method (Balasubramaniam et al., 2015). It is also used as a means of inactivating pathogens within packaged ready-to-eat meats and to inactivate spoilage enzymes of avocados and guacamole to make them more stable under refrigeration.

HPP has a unique application for the shellfish industry since it can separate the adductor muscle from the oyster shell, effectively shucking shellfish (Martin and Hall, 2006). Also HPP is used as an intervention to inactivate pathogenic *Vibrio vulnificus* (Vv) in oysters grown in waters where this bacteria is endemic, reducing *Vv* to "nondetectable levels." Recent studies indicate that pressure alone or in combination with mild heat or cold storage eliminates *V. parahaemolyticus* in oysters as well (Ye at al., 2012, 2013).

Commercial HPP units can have capacities exceeding several hundred liters. HPP is a batch process with machines filled and pressurized for short intervals of < 5 min, followed by depressurization and removal of the treated food item. Commercial HPP units are usually water-based but research units can use water, oil, alcohol or a combination of these as the pressurization media. Time to reach the desired pressure is commonly termed ''come-up'' time. This time varies with different types and sizes of HPP units and, depending on the type of pressure unit, can contribute to the inactivation observed. For most machines, pressure release is very quick, being achieved in a few seconds, if not instantly. One challenge for widespread commercial HPP is that it is generally ineffective against bacterial spores (Akhtar et al., 2009; Shearer et al., 2000). As a result, HPP is generally only performed on refrigerated foods, or in situations where bacterial spores are not a concern.

Prior to the last decade and a half, most HPP studies focused on inactivating foodborne vegetative bacteria, fungi, and protozoa. In general, bacterial vegetative cells, yeasts, and molds can be inactivated by pressure applications ranging from 100 to 800 MegaPascals (MPa; Balasubramaniam et al., 2015). One MPa is approximately equivalent to slightly less than 10 atmospheres of pressure or slightly less than 150 pounds per square inch of pressure. For these organisms, HPP is principally thought to reversibly disrupt cellular membrane integrity, causing leakage of cytosolic contents (Wang et al., 2016). Research since the year 2000 has shown that HPP can inactivate foodborne viruses. Many human and animal viruses are inactivated at pressures ≤ 600 MPa, presumably by disruption of protein structure since most foodborne viruses do not have membranes (lipid envelopes). Since HPP does not disrupt covalent bonds, smaller molecules such as vitamin C and ß-carotene are unaffected (Balasubramaniam et al., 2015)

HPP Principles

Biochemical and physicochemical processes are often in equilibrium. Changes in pressure will cause changes in volume. Reducing volume on a molecular level will shift equilibria in favor of reducing volume and disfavor increases in molecular volume. Thus, HPP enhances chemical reactions that favor a volume decrease such as weak acids where ionic dissociation is favored, making weak acids stronger under pressure (Kitamura and Itoh, 1987). Generally speaking, covalent bonds are not broken or altered by pressure (Wang et al., 2016). Hydrogen bonds, which substantially influence protein structure and polypeptide interactions, tend to be stabilized at relatively low pressures, but can be perturbed at more extreme pressures (Hendrickx et al., 1998; Knorr, 1999, Balasubramaniam et al., 2015).

Protein structure is commonly described in terms of primary, secondary, tertiary and quaternary structures. The primary structure of a protein is basically its sequential composition of individual amino acids. Secondary structure is the structure of the amino acid strand such as helix, beta sheet or random coil. Tertiary is how that strand is positioned in 3-dimensional space, and quaternary is how one protein subunit is positioned relative to another protein or subunit.

The secondary structure of proteins, determined mainly by multiple hydrogen bonds, is only disrupted at very high pressures, leading to irreversible denaturation (Hendrickx et al., 1998; Knorr, 1999). Depending on the barostability of an individual protein, the tertiary structure of proteins, which is maintained principally by hydrophobic and ionic interactions, can be disrupted at pressures above 200 MPa (Hendrickx et al., 1998). Also for complex proteins, the quaternary structures can be subtly altered resulting in a disruption of function, perhaps denaturing an enzyme, or altering the virus capsid slightly, blocking its ability to initiate an infection.

Pressure is isostatic, meaning it is instantaneous and uniform. Pressure is immediately transmitted throughout food systems regardless of the products' shape, size or geometry. For HPP, pressurization cycles are rapid, allowing shorter processing times relative to thermal processes, such as canning (Balasubramaniam et al., 2015). HPP systems are energy efficient, since once the required pressure is reached; it can be maintained without additional energy (Balasubramaniam et al., 2015; Dumay et al., 2006).

Formally-speaking HPP is a non-thermal technology. However HPP does alter temperatures somewhat. When pressure is applied, compressive work is performed against intermolecular forces, causing an effect called "adiabatic heating". The actual temperature increase in HPP-treated food is influenced by food composition, starting temperatures, pressurization rate, the size of the processing equipment and, most importantly the composition of the pressure-transmitting media, which is normally water in commercial situations (Otero et al., 2007). Most foods follow compressive heating behavior that is similar to water, with an adiabatic heating value of approximately 3-4°C/100 MPa (Rasanayagam et al., 2003). Fats and oils, on the other hand, have high compression heating values (8 to 9°C per 100 MPa). Some commercial hybrid thermal HPP processes have been designed to exploit compression heating by preheating to a somewhat high temperature, then applying pressure to achieve the desired thermal lethality threshold (Balasubramaniam et al., 2015; Wilson et al., 2008). Pressurization rate will determine how much time is permitted for adiabatic heat to dissipate to the environment. Thus slow pump rate pressure units will have less adiabatic heating than faster units. Typically commercial HPP units are large, having a relatively high volume to surface ratio, causing heat to dissipate more slowly. Smaller research units have small volumes and greater surface area ratios so they dissipate adiabatic heat to the environment more quickly. When pressure is released, the reverse effect happens, termed "adiabatic cooling". This temperature decline is very rapid and, depending on how much adiabatic heat was dissipated to the environment during pressurization, can result in food temperatures below the original starting temperature.

Mechanisms of HPP Inactivation for Viruses

The mechanisms of inactivation by HPP are thought to involve alteration of virus protein conformation and not to inactivate microorganisms by damage to genetic material. Research indicates that pressure-induced changes to the viral coat can cause alterations to capsid proteins or receptor-recognition proteins to prevent virus attachment, penetration of the host cell, or uncoating after entering the cell. For example, after inactivation of HAV at 500 MPa, the capsid still remains intact, since the RNA genome was still protectected from degradation by RNAse (Kingsley et al., 2002). In an analogous manner, HPP was shown to cause limited disruption of rotavirus capsids, and reduced infectivity is attributed to hemagglutinin (VP4) damage, which forms spikes in the outer shell and is involved in receptor recognition (Pontes et al., 2001). For MNV, a high pressure treatment for 5-min at 400-MPa and 0°C abrogated host cell binding ability, indicating a loss of receptor-binding function by the virus capsid (Tang et al., 2010). Tang et al. (2010) also noted that MNV RNA was protected from RNase treatment after the 400-MPa treatment, which indicated that the MNV capsid remained intact. Lou et al. (2011) also evaluated HPP's mechanism of inactivation for MNV. After applying 350 MPa of pressure for 2 min and visualizing MNV-1 by EM, the virion structure was found to be altered from a discrete small round shape to undefined protein debris. However, the individual capsid proteins of MNV-1 were not degraded since they still bind MNV-1 antibody, suggesting that the primary and secondary structures of viral proteins were not disrupted. A pressure treatment of 600 MPa destroyed MNV integrity, causing loss of the viral RNA genome (Lou et al., 2011).

Barosensitivity of Human Norovirus and Surrogates

Initial work attempting to assess the potential of HPP to inactivate HuNoV was performed using the propagable surrogate viruses, feline calicivirus (FCV) and murine norovirus (MNV). The first indication that HPP might be capable of inactivating HuNoV was the evaluation of FCV when 5 min treatments of 250 MPa at room temperature were sufficient to inactivate 7-log of FCV in culture media (Kingsley et al., 2002). Later, Grove et al. (2006) demonstrated that FCV viability was reduced by 3.6-log after a 2 min treatment with 300 MPa and undetectable after 3 min-300 MPa. In oysters, Murchie et al. (2007) demonstrated that a pressure level of 300 MPa for 5 min reduces FCV to undetectable levels (>3.8 log reduction).

Subsequently, MNV was discovered (Karst et al., 2003; Wobus et al., 2006) and utilized in HPP research. Based on its ability to infect mice orally, to replicate in the murine gastrointestinal tract, and its genetic classification as a norovirus, the mouse virus was thought to be a superior surrogate. FCV infects felines via a respiratory route and, while classified as a calicivirus, is not a member of the norovirus family. Initial evaluation of MNV stocks in DMEM tissue culture media showed that while MNV was sensitive to pressure, it was more resistant to pressure than FCV. For example, a 5-min 450-MPa treatment was sufficient to inactivate 6.9-log of MNV using the Avure PT-1 unit (Kingsley et al., 2007).

Farkas et al. (2008) reported the discovery of a propagable primate norovirus named the Tulane virus (TV), after it was isolated from the Tulane National Primate Research Center. TV has a genomic sequence that resembles HuNoV and more importantly, it recognizes the histo-blood group antigens (HBGAs). HBGAs are virus receptors required for infection of the gastrointestinal tract by HuNoV, suggesting that TV's infectious process is highly similar to HuNoV's (Farkas et al., 2010). Comparison of

MNV and TV indicated that a 2 min-350 MPa treatment at room temperature inactivated ~ 4 log in culture, indicating that TV was more sensitive to pressure compared to MNV (Li et al., 2013b; Li and Chen, 2015).

It is important to recognize that the interpretation of inactivation study results using surrogates may or may not be applicable to HuNoV. Different taxonomically-related viruses behave differently under pressure, as exhibited by the difference between FCV, MNV and TV and the range of pressure sensitivities observed for the picornaviruses. Indeed the limited information about HuNoV under pressure suggests that a GI.1 strain is more baroresilient than a GII.4 strain (Huang et al., 2016: Lou et al., 2015; Sido et al., 2016) and some GII strains are more resistant than others (Lou et al., 2016). Thus assumptions that a surrogate will behave in a manner analogous to the human pathogenic viruses may not be valid.

To directly evaluate the potential of HuNoV to be inactivated by HPP, a clinical trial was performed. Human subjects were fed pressure-treated and untreated oysters inoculated with HuNoV GI.1 via injection (Leon et al., 2011). Five min-600 MPa HPP treatment of oysters at 6°C resulted in no infection among subjects (0 out of 10 subjects infected) while 5 min-400 MPa pressure treatments at 6°C reduced the infection rate to 21% (3 out of 14 subjects infected) whereas 7 of 15 volunteers became infected when given untreated virus.

As noted above, human norovirus is difficult to propagate in the laboratory. For HuNoVs, RT-PCR assay is currently the standard detection method since it is highly sensitive and rapid. Unfortunately RT-PCR only detects HuNoV RNA, not capsid damage and, therefore, it cannot easily distinguish between infectious and non-infectious viral particles. As a result, the amount of infectious norovirus compared to inactivated norovirus in a given food is unknown (Baert et al., 2008; Lowther et al., 2010). This is also an important caveat when considering virus inactivation by HPP. As most food-based testing only determines the presence of the viral genome, and HPP targets the virus capsid and not the virus genome, foods may still test positive for the presence of the virus, despite being inactivated.

Porcine gastric mucin (PGM) binds to genogroup I (GI) and genogroup II (GII) HuNoVs (Tian et al., 2008; Tian et al., 2010). This feature was utilized to extract and concentrate HuNoVs from foods by means of PGM-conjugated magnetic beads (PGM-MBs) (Tian et al., 2008; Tian et al., 2010). Subsequently this assay was shown by Dancho et al. (2012) to be a means to separate infectious and non-infectious HuNoV based on the assumption that only the infectious virus could bind to PGM. The separated virus particles can then be quantitated by RT-qPCR. Dancho et al. (2012) reported that application of 5-min 600-MPa HPP treatments to GI.1 HuNoV reduced binding to PGM by 4.7-logs, while a 300-MPa pressure treatment, reduced binding to PGM by only 0.5-logs and a >3-log reduction in binding was noted for a 400 MPa treatment. Overall these results appeared consistent with the clinical trial performed by Leon et al. (2011). Li and Chen (2015) compared HPP inactivation results of MNV-1, TV and HuNoVs obtained through the PGM-MB binding assay and concluded that the PGM-MB binding assay quantifies HPP inactivation of HuNoV.

Using the PGM-MB binding assay and gnotobiotic swine *in vivo* to assess HuNoV viability, Lou et al. (2015) showed that 350 MPa applied at 0°C for 2 min led to a 3.7 log reduction in viral RNA copies in oysters homogenate, while newborn gnotobiotic piglets orally-fed oyster homogenate treated at the same pressure level did not shed the virus in their feces, or show evidence of viral replication in the small intestine. These results indicate that HuNoV survival estimated by an *in vitro* PGM-MB viral binding assay are consistent with the infectivity determined by an *in vivo* gnotobiotic pig model and, importantly, that HPP is capable of inactivating the predominate HuNoV GII.4 strain at commercially-viable pressure levels.

Barosensitivity of Other Foodborne Viruses

Using the tissue culture-adapted HM-175 hepatitis A virus strain, HPP inactivation of the virus was first demonstrated utilizing an oil-based custom-built HPP unit, where pressure treatments were performed at room temperature (Kingsley et al., 2002). Pressure had no discernible effect on the virus (HAV stock in DMEM tissue culture media) until pressures >300 MPa were used. However after application of 460 MPa, complete inactivation of the 7-log HAV stock was observed. Mixing HAV with seawater to achieve approximately 2.5% salinity substantially reduced HPP inactivation (Kingsley et al., 2002). To investigate the role of time under pressure inactivation, treatment time was extended from 5 to 15 min. This did enhance observed inactivation somewhat, but overall the pressure level applied during HPP had much more influence on the inactivation than extending the time of pressurization. Grove et al. (2008) also independently evaluated HAV, reporting that 10 min HPP at 300, 400, and 500 MPa showed >1, > 2, and > 3.5 log $TCID_{50}$ reductions, respectively. Presently, the degree to which HuNoVs and wild-type HAV strains may vary in pressure sensitivities is unknown due to in vitro propagation difficulties. Almost all HPP inactivation research has also been with one strain of tissue culture-adapted HAV, which is presumably no longer pathogenic. Shimasaki et al. (2009) did evaluate four different strains of HAV for barosensitivity, finding one strain to be more sensitive to HPP while the other three strains had comparable sensitivity to the tissue culture-adapted HM-175 strain described above.

Other potential foodborne viruses include the human picornaviruses. Pressure treatments have been experimentally performed on coxsackie A9, coxsackie B5, polio, aichivirus, and HPeV. Collectively speaking, variable pressure sensitivities were noted for these different viruses. Some viruses proved to be highly resistant, while others had limited resistance, but higher resistance than HuNoV and HAV. Wilkinson et al. (2001) originally noted resistance of poliovirus to HPP to 1 hr-600 MPa treatments. Subsequent work showed that coxsackie B5, polio (Chat strain), and aichivirus were found to be completely resistant to 5-min 600-MPa pressure treatments, when pressure-treated at room temperature using a custom-built oil based pressure unit (Kingsley et al., 2002; 2004). Coxsackie A9 was somewhat sensitive to HPP with a 5 min-400 MPa treatment at room temperature, inactivating 3.4 log tissue culture infectious dose 50% ($TCID_{50}$). For HPeV, a 5-min 400-MPa treatment at room temperature resulted in a 1.3-log reduction, while a 5-min 500-MPa treatment at room temperature inactivated 4.3- $TCID_{50}$ (Kingsley et al., 2004). Rotavirus is thought to be sensitive to HPP. A study by Khadre and Yousef (2002) demonstrated a substantial (8-log) reduction after a 2-min 300-MPa treatment at room temperature, although a small proportion of the rotavirus was noted to be highly resistant to pressures. Subsequent work by Araud et al. (2015) evaluating seven rotovirus strains belonging to four rotavirus serotypes indicated that GI serotype strains were most resistant and G4 strains were the most sensitive to pressure. Araud et al. (2015) noted that a 400 MPa treatment for 2 min was sufficient to inactivate the GI serotype Wa strain, the most pressure-resistant strain. For adenoviruses, treatments above 400 MPa at room temperature inactivate type D and AdV2 (Kovač et al., 2012; Wilkinson et al., 2001). Pressure sensitivity of HEV and sapovirus to HPP are currently unknown.

APPLICATION OF HPP TO FOODS

Bivalve Mollusks

Anthropogenic and indigenous environmental microbial contamination are chronic and pervasive in shellfish growing and harvesting areas, making raw shellfish a high risk food. A couple of factors account for bivalve mollusks as high risk foods. Shellfish are filter feeders that selectively concentrate particles of phytoplankton, zooplankton, bacteria and viruses from the surrounding waters by a factor of as much as 1000-fold (Canzonier, 1971). As a result, bacteria from human waste, such as *Salmonella spp.* can bioaccumulate within bivalve shellfish. *Vibrio vulnificus* and *Vibrio parahaemolyticus* bacteria are indigenous to brackish tropical estuaries and frequently found in temperate climates during the summer season when water temperatures are elevated. When shellfish are harvested from warm areas, vibrios can replicate within shellfish at very high levels (Kingsley, 2014b). Infectious viruses from human waste also readily bioconcentrate within shellfish. As mentioned earlier, they do not replicate within bivalves but viruses can be found in high numbers in human waste and they are probably substantially resistant to sewage treatment. Also phytoplankton and dinoflagellate blooms often produce biotoxins which can reach elevated concentrations within shellfish. Compounding this bioconcentration issue is the fact that shellfish are frequently eaten raw, or perhaps lightly-cooked (Lees, 2000), permitting any pathogens to remain infectious.

HPP has the potential to make a revolutionary change in the shellfish industry. It can greatly facilitate the shellfish shucking process and does not detrimentally affect the organoleptic properties of raw oysters (Cruz-Romero et al., 2004; Hayashi, 1992; Kingsley et al., 2014a; Lopez-Caballero et al., 2000; Murchie et al., 2005). Shucking occurs at approximately 225 MPa, or about 33,000 psi (Kingsley, unpublished results) and HPP treatments do not significantly change color or texture of oyster tissue (Ye at al., 2015). Current commercial treatments are approximately 275 MPa to about 300 MPa for >5 minutes at room temperature. This is sufficient to reduce *V. vulnificus* to "nondetectable levels" within oysters, eliminating risk for susceptible oyster consumers. Pressures of 300-325 MPa, that are only slightly higher than current commercial pressures, are needed to inactivate *V. parahaemolyticus* (Kural and Chen, 2008; Kural et al., 2008; Ye at al., 2012, 2013).

Direct testing of virus inactivation within live shellfish contaminated by exposure to contaminated water or testing in the context of oyster homogenates is critical, since the food matrix can potentially reduce the efficiency of HPP. Within oysters, Kingsley et al. (2007) observed a 4-log decrease of MNV-1 after pressure treatment of oysters for 5 min at 400 MPa (Kingsley et al., 2007). Li et al. (2009) subsequently demonstrated that a 5-min 400-MPa treatment at 0°C inactivated MNV-1 in oysters to undetectable levels. Experiments have shown that, for MNV at least, high pressure inactivation was equal when assessed by *in vivo* infection of mice or *in vitro* cell culture (Gogal et al., 2011). The clinical trial reported by Leon et al. (2011) is the only study where the pressure inactivation profile of GI.1 HuNoV has been directly tested. As previously described, human volunteers were fed pressure-treated and un-treated oysters injected with HuNoV GI.1. There was no statistical difference between the HuNoV infection rates of subjects fed oysters treated at 400 MPa for 5 min at 25°C (3 out of 5 or 60% of subjects infected) and subjects fed untreated oysters (7 out of 15 or 47% of subjects infected). HPP treatment (400 MPa for 5 min) conducted at 6°C reduced the infection rate to 21% (3 out of 14 subjects infected). None of the 10 subjects fed with HuNoV-seeded oysters treated by 600 MPa at 6°C for 5 min became infected, demonstrating a > 4 log reduction of genomic equivalent copies of HuNoV. Ye et al. (2014) studied the effect of HPP on inactivation of HuNoV using the PGM-MB binding method (pretreatment of samples with RNase, extraction of viral particles using PGM-MB binding, and quantification of viral RNA using RT-PCR) used for quantifying infectious HuNoV within the context of shellfish homogenates. A 450 MPa treatment at 1°C achieved a > 4 log reduction of GI.1 HuNoV in both oyster and clam homogenates and the more common GII.4 strain was even more sensitive to HPP than the GI.1 strain. These results correlate well with observations from the human challenge study (Leon et al., 2011) and indicate that the PGM-MB binding method reflects the infectivity of HuNoV after HPP. Ye et al. (2015) also recently reported that HPP at 350 and 500 MPa at 0°C inactivates >4-logs of GII.4 and GI.1 HuNoV within oyster homogenates, respectively.

The first work evaluating the potential of HPP to inactivate HAV within shellfish was performed using live oysters contaminated with up to 6-log of virus after exposure to contaminated seawater. Pressure treatments were performed using a water-based Quintas QFP-6 (ABB Autoclave Systems, Inc., Columbus, OH) for 1 min at 9°C. Inactivation observed was >1, >2, and >3 log of HAV, after treatment with 350, 375, and 400 MPa respectively (Calci et al., 2005). Although pressure is uniform when applied and would not be expected to be influenced by the presence or absence of a bivalve shell, inactivation of HAV within "in shell" oysters, which mimics commercial conditions, and within shucked oysters (research conditions) was evaluated. Results showed that in-shell and shucked-oysters had essentially equivalent inactivation (Kingsley et al.,

2009a). HPP has also been shown to inactivate HAV within mussels. HAV was bio-accumulated within live Mediterranean (*Mytilus galloprovincialis*) and blue mussels (*Mytilus edulis*), followed by treatment at 350 and 400 MPa for 5 min at room temperature. Reductions of 1.7 and 2.9, 2.1 and 3.6 logs of HAV were observed within Mediterranean and blue mussels, respectively (Terio et al., 2010). Arcangeli et al. (2012) evaluated MNV within clams and found that 500 MPa for 1-min was sufficient to inactivate the virus.

HPP does have some drawbacks and negative perceptions by shellfish consumers and industry. First, as with thermal methods, HPP cannot inactivate biotoxins. Second, when HPP is performed commercially, it is typically performed to 275-300 MPa and at room temperature. Inactivation of HuNoV and HAV by HPP will require pressures substantially higher than 300 MPa. Those utilizing HPP on shellfish have argued that pressures above 300 MPa gives an inferior product. In addition, extemporaneous conversations with consumers and retailers of HPP-treated oysters indicate that these shellfish may not be as desirable as fresh-shucked oysters. Lastly in the US, there are currently only a few companies that utilize HPP technology for shellfish. This is probably due to equipment expense. Basic commercial equipment costs several hundred thousand dollars. Ostensibly, this means that high product throughput is required to take advantage of economies of scale. Thus unless you are an exceptionally large shellfish producer, able to utilize an HPP machine around the clock, some form of a consortium would be needed for multiple smaller shellfish producers to utilize this technology.

To address the overall question of quality of HPP-treated oysters and the question of whether oysters treated at pressures capable of inactivating HuNoV and HAV would be acceptable, volunteers who professed to enjoy eating raw oysters were utilized for a consumer preference study. Flavor, aroma, appearance, texture, and general acceptability of pressure-treated oysters were evaluated. Treatments of 300 MPa, 400 MPa and 500 MPa at room temp (22°C) for 5 min and 400, 500 and 600 MPa treatments at 6°C for 5 min, were performed, respectively. This study found that all one day-old HPP-treated oysters were actually slightly preferred to fresh-shucked oysters (Kingsley et al., 2015). Thus HPP treatment performed at $\geq$ 400 MPa, that would be needed to inactivate shellfish-borne HAV and HuNoV within shellfish meat, are more than acceptable to the consumer.

So why are there negative perceptions about HPP-treated shellfish? In this author's opinion, the perception stems from the need to process oysters in a central high throughput location, followed by shipping and distribution. This can result in a half-shell HPP oyster being consumed as much as a week after processing. While research has demonstrated that HPP can reduce spoilage organisms on, and within, raw shellfish and extend refrigerated shelf-life somewhat (He et al., 2002; Kingsley 2014a; 2014b), it stands to reason that a fresh-shucked oyster would be preferred to a week-old HPP-treated one. Thus the critical challenge for HPP will be to provide freshly-processed oysters. Under pressure, shellfish can absorb fluid which increases the weight of oyster meat. This was recently utilized to flavor an oyster by driving sriracha sauce flavoring directly into the meat, demonstrating that HPP could be used to make unique shellfish food products (Kingsley et al., 2015).

Fruits and Vegetables

Hand-picked fruits and vegetables offer the potential for soiled fingers to contact the contaminate produce (Baert et al., 2011). Crop irrigation, or fertilizer and pesticide applications, with non-potable water that has been contaminated by human waste is another important contamination source of these viruses (Hall et al., 2012). In many cases, HPP can effectively inactivate surface and internal viral contaminants, while preserving key organoleptic food characteristics. That said, plant tissues that contain air will be compressed during pressure treatment, inducing physical damage to the tissues (Michel and Autio, 2002). Therefore HPP is best for products that use fruits and vegetables as flavor enhancers, or as ingredients for frozen desserts, jams, juices, and smoothies. Guacamole, fruit jellies, salsa, chopped Spanish onions, green peppers, orange juice and apple sauce are among some of the commercial food applications of HPP. HPP has also been applied commercially to chopped fruit salad-type products.

Initial proof-of-principle for the inactivation for viruses within produce was performed more than 10 years ago using the Avure PT-1 high pressure unit (Avure Technologies, Inc., Kent, WA). A 5 min-350-MPa treatment at 21°C was able to inactivate 4-log of HAV within strawberry puree and within sliced green onions (Kingsley et al., 2005). It was noted that HAV within strawberry puree was more sensitive to HPP than virus stocks in DMEM tissue culture media or within green onions. This noted difference is now attributed to the acidic nature of strawberries (Kingsley and Chen, 2009b). Subsequent work showed that 2-min 350-MPa treatments at 4°C resulted in 2.2-, and 2.4-log reductions of MNV-1 on fresh-cut strawberries, and within strawberry puree, respectively, while product quality was minimally impacted (Lou et al., 2011). Li et al. (2013b) demonstrated that MNV-1 and TV on blueberries were inactivated by pressures of $\leq$ 400 MPa. For strawberries, HPP at 1 min-350 MPa treatments at 0°C achieved > 4 log reductions of MNV-1 in both strawberry puree and cut strawberries (Huang et al., 2014). A treatment of 500 MPa for 5 min at 20°C effectively inactivated internalized MNV-1, HAV, and adenovirus type 41 in green onions (Hirneisen and Kniel, 2013).

HuNoV has been evaluated in the context of sliced strawberries, sliced raspberries, whole blueberries and their purees. Huang et al. (2016) report that a 2 min-550 MPa treatment at 0°C eliminates >2.9 log of GI.1 and >4 logs of GII.4 HuNoV, respectively, within these purees. Sido et al. (2016) evaluated MNV and HuNoV in salsa and green onions, reporting that 600 MPa and 500 MPa treatments at 1°C for 2 min inactivates > 3 logs of GI.1 HuNoV within green onions and salsa, respectively. For a GII.4 HuNoV strain in green onions and salsa, 500 MPa and 300 MPa at 1°C for 2 min, respectively, gave a > 3 log reduction (Sido et al., 2016).

CRITICAL FACTORS FOR HPP

Research has shown that HPP processing parameters such as pressure magnitude, time under pressure and initial pressurization temperature, can substantially affect the inactivation of microorganisms within foods. In general, increasing applied pressure levels and time under pressure results in increased inactivation. However for economic reasons, shorter treatment times and lower pressure levels that are easier on equipment are generally preferred by industry. Extending the time of pressure application results in increased inactivation for viruses evaluated, such as FCV, MNV, and HAV, but that increase diminishes with time (Kingsley, 2013). As an illustration, a 20 min-200 MPa treatment at room temperature reduced the titer of FCV by 2.8 log, while extending the treatment time to 76 min only increased the level of reduction by an additional 0.9 log (Chen et al., 2005). Inactivation curves of pressure applied versus variable time applied fit most closely with Weibull and log-logistic curves (Chen et al. 2005; Kingsley et al. 2006; 2007), as compared to first-order curves. The observed tailing of inactivation curves indicates that virus inactivation is more dependent on the level of pressure applied than treatment time. Curiously, oscillatory high pressure processing did not considerably enhance pressure inactivation of HAV when 2, 4, 6, and 8 cycles from 0 to 400 MPa were compared with continuous high pressure (Kingsley et al. 2006).

The temperature at which initial pressure is applied can have a profound effect on inactivation. It is known that the dissociation and denaturation of proteins of viruses by pressure can be promoted by low temperatures (Kingsley and Chen, 2008; Kunugi and Tanaka, 2002; Tian et al., 2000). The HuNoV and surrogates, FCV, MNV-1, and TV, are all significantly more sensitive to HPP at lower temperatures (Chen et al., 2005; Kingsley et al., 2007; Li et al., 2013b). MNV-1 is an example. Five-min-350 MPa treatments of MNV-1 at 5°C resulted in a 5.6 log reduction (Kingsley et al., 2007), while 20 and 30°C resulted in only 1.8 and 1.2 log reductions, respectively. For both GI.1 and GII.4 HuNoV, inactivation increased with decreasing initial sample temperature (1 > 4 > 10 > 21 > 35°C; Li et al., 2013b).

Other viruses are not so cold-sensitive. Two studies that highlight this fact were performed by Kingsley et al., (2006) and Kingsley and Chen (2009b). In those studies, temperature effects on HPP inactivation of HAV was actually reduced by colder initial HPP treatment temperatures. This is a stark contrast to HuNoV. For other picornaviruses, the initial pressurization temperature results are variable. Pressure inactivation of coxsackie A9 and coxsackie B5 viruses is enhanced, while HPeV has reduced inactivation at refrigeration temperatures (Kingsley et al., 2014b). For rotavirus, Araud et al. (2015) found that 6 of 7 strains tested were more sensitive to pressure at 4ºC than 20ºC, while one strain showed no temperature differences. Also avian influenza virus, an enveloped virus, was found to be more sensitive to HPP as temperatures increase from 15 to 30°C (Isbarn et al. 2007).

The matrix composition or food within which the virus is sequestered can have a substantial effect on the response of virus to pressure. Certain food constituents, such as fats, proteins, salts and sugars, can reduce HPP effectiveness and increase virus resistance to pressure (Baert et al. 2009; Balny et al. 2002, Gross and Jaenicke, 1994, Kingsley and Chen, 2009b).

Matrix pH is also a key consideration. FCV, MNV-1, TV, HuNoV and adenovirus are all more sensitive to HPP at neutral pH than at acidic pH (Chen et al., 2005; Kingsley and Chen, 2008; Kovăc et al., 2012; Li et al., 2013a; Lou et al., 2011). At pH 7, treatments of 300 MPa at 4°C for 2 min reduced MNV-1 by 5.5 logs, whereas at pH 4 only a 2.5-log virus reduction was achieved under the same conditions (Lou et al. 2011). Using the PGM-MB receptor binding assay, Li et al. (2013a) showed both GI.1 and GII.4 HuNoVs to be more sensitive to pressure at pH 7.0 than at pH 4.0. In contrast, HAV was found to be more sensitive to pressure in an acidic environment over neutral pH despite being very stable at low pH under normal atmospheric pressure. A decline in pH value from neutral pH to an acidic pH of 3 was shown to increase the pressure inactivation of HAV by 3 log PFU/ml, following treatment at 400 MPa for 1 min at 20°C (Kingsley and Chen, 2009b).

The salinity of the matrix is a key consideration. The mechanism underlying the protective effect from salt is not well understood. One possibility is that salt may help stabilize the solvation cage surrounding the hydrophilic virus proteins and maintain the void volumes by displacing water molecules, effectively altering solution density and compressibility. Kingsley and Chen (2008) conducted a pressure inactivation study of FCV exposed to moderate and high concentrations of NaCl. A baroprotective effect was observed for NaCl at concentrations up to 12%, but protection plateaued at NaCl concentrations >12%. Reduction of HPP inactivation by NaCl was also demonstrated for HAV (Kingsley and Chen, 2009b), where 400 MPa-1 min at 50°C gave a 4.0-log reduction without NaCl, while 3% and 6% NaCl reduced the HAV titers in an aqueous medium to 1.3 and 0.4 log PFU/ml, respectively. Also HAV in seawater required higher pressures for comparable levels of inactivation, as observed in isotonic media (Kingsley et al., 2002). A similar reduction of HPP efficiency was observed for sucrose when concentration was as high as 40%. When both NaCl and sucrose are added, an additive inhibitory effect on FCV was observed (Kingsley and Chen, 2008). While HPP is a water-dependent technology, this solute effect is not simply a function of water activity, since it was noted that FCV samples with identical water activity containing either enhanced amounts of NaCl or sucrose had differing amounts of inhibition (Kingsley and Chen, 2008).

Foods, of course, are complex matrices that can inhibit HPP inactivation of viruses. While the effects of pH, salt and sugar can serve as guidelines, direct testing for the pathogen of interest within the food product matrix will be needed to design specific HPP interventions. Research described here should be a basic starting point to evaluate whether a current or proposed commercial process could be effective against HuNoV or other viruses. Given the complexities of food matrices, different viruses and perhaps even different virus strains can have variable responses to pressure. This information can serve as a means of estimating the potential effects of changes in food formulations. Finally, it should be noted that an intervention such as HPP should be utilized as a fail-safe, or final assurance of food safety, rather than a means to remediate unsanitary foods.

Future Challenges and Directions

The direct determination of HuNoV inactivation during HPP remains difficult. The PGM-MB assay, human volunteers, and gnotobiotic swine have provided critical information about pressures needed to inactivate HuNoV. However all of these methods have drawbacks, either not directly demonstrating inactivation, as is the case for the PGM-MB assay, or they are expensive and logistically-complex, such as human volunteers and, to a lesser degree gnotobiotic swine. Recent research publications are now reporting some limited success using direct propagation of HuNoV (Ettayebi et al., 2016; Jones et al., 2015). For example, the method described by Ettayebi et al., (2016) utilizes the 3-dimensional enteroid tissue culture method. At present, small quantities of slow-growing enteriod cells that were originally derived from human patients are grown in a collegen or gel-type matrix and then seeded onto plates in monolayers where end-stage differentiation provides limited replication generating a 2- or 3-log increase of virus over inoculation levels. It is hoped that future improvements to this system will make this propagation system more practical for routine testing.

References

Akhtar, S., Paredes-Sabja, D., Torres, J. A. & Sarker, M. R. (2009). Strategy to inactivate *Clostridium perfringens* spores in meat products. *Food Microbiol*, *26*, 272-277.

Amon, J. J., Devasia, R., Xia, G., Nainan, O. V., Hall, S., Lawson, B., Wolthuis, J. S., Macdonald, P. D., Shepard, C. W., Williams, I. T., Armstrong, G. L., Gabel, J. A., Erwin, P., Sheeler, L., Kuhnert, W., Patel, P., Vaughan, G., Weltman, A., Craig, A. S., Bell, B. P. & Fiore, A. (2005). Molecular epidemiology of foodborne hepatitis A outbreaks in the United States, 2003. *J Infect Dis*, *192*, 1323-1330.

Anon. (2003). Hepatitis A outbreak associated with green onions at a restaurant--Monaca, Pennsylvania, 2003, *MMWR*, *52*, 1155-1157.

Araud, E., DiCaprio, E., Yang, Z., Li, X., Lou, F., Hughes, J. H., Chen, H. & Li, J. (2015). High-pressure inactivation of rotaviruses: Role of treatment temperature and strain diversity in virus inactivation. *Appl Environ Microbiol*, *81*, 6669-6678.

Arcangeli, G., Terregino, C., De Benedictis, P., Zecchin, B., Manfrin, A., Rossetti, E., Magnabosco, C., Mancin, M. & Brutti, A. (2012). Effect of high hydrostatic pressure on murine norovirus in Manila clams. *Lett Appl Microbiol*, *54*, 325-329.

Atmar, R. L. & Estes, M. K. (2006). The epidemiologic and clinical importance of norovirus infection. *Gastroenterol Clinics*, *35*, 275-290.

Baert, L., Debevere, J. & Uyttendaele, M. (2009). The efficacy of preservation methods to inactivate foodborne viruses. *Int J Food Microbiol*, *131*, 83-94.

Baert, L., Mattison, K., Loisy-Hamon, F., Harlow, J., Martyres, A., Lebeau, B., Stals, A., Van Coillie, E., Herman, L. & Uyttendaele, M. (2011). Review: Norovirus prevalence in Belgian, Canadian and French fresh produce: A threat to human health? *Int J Food Microbiol*, *151*, 261-269.

Baert, L., Wobus, C. E., Van Coillie, E., Thackray, L. B., Debevere, J. & Uyttendaele, M. (2008). Detection of murine norovirus-1 by using plaque assay, transfection assay, and real-time reverse transcription-PCR before and after heat exposure. *Appl Environ Microbiol*, *74*, 543-546.

Balasubramaniam, V. M., Martínez-Monteagudo, S. I. & Gupta, R. (2015). Principles and application of high pressure-based technologies in the food industry. *Ann Rev Food Sci Technol*, *6*, 435-462.

Balny, C., Masson, P. & Heremans, K. (2002). High pressure effects on biological macromolecules: from structural changes to alteration of cellular processes. *Biochim Biophys Acta, Prot Struct Mol Enzymol*, *1595*, 3-10.

Bouwknegt, M., Verhaelen, K., Rzeżutka, A., Kozyra, I., Maunula, L., von Bonsdorff, C. H., Vantarakis, A., Kokkinos, P., Petrovic, T., Lazic, S., Pavlik, I., Vasickova, P., Willems, K. A., Havelaar, A. H., Rutjes, S. A. & de Roda Husman, A. M. (2015). Quantitative farm-to-fork risk assessment model for norovirus and hepatitis A virus in European leafy green vegetable and berry fruit supply chains. *Int J Food Microbiol*, *198*, 50-58.

Burkhardt, W., Blackstone, G., Skilling, D. & Smith, A. (2002). Applied technique for increasing calicivirus detection in shellfish extracts. *J Appl Microbiol*, *93*, 235-240.

Calci, K. R., Meade, G. K., Tezloff, R. C. & Kingsley, D. H. (2005). High-pressure inactivation of hepatitis A virus within oysters. *Appl Environ Microbiol*, *71*, 339-343.

Cannon, J. L., Papafragkou, E., Park, G. W., Osborne, J., Jaykus, L. & Vinje, J. (2006). Surrogates for the study of norovirus stability and inactivation in the environment: A comparison of murine norovirus and feline calicivirus. *J Food Protect*, *69*, 2761-2765.

Canzonier, W. J. (1971). Accumulation and elimination of coliphage S-13 by the hard clam, *Mercenaria mercenaria*. *Appl Microbiol*, *21*, 1024-1031.

Carter, M. (2005). Enterically infecting viruses: pathogenicity, transmission and significance for food and waterborne infection. *J Appl Microbiol*, *98*, 1354-1380.

Carvalho, C., Thomas, H., Balogun, K., Tedder, R., Pebody, R., Ramsay, M. & Ngui, S. (2012). A possible outbreak of hepatitis A associated with semi-dried tomatoes, England, July-November 2011. *Euro Surveill*, *17*, 1-4.

Cheetham, S., Souza, M., Meulia, T., Grimes, S., Han, M. G. & Saif, L. J. (2006). Pathogenesis of a genogroup II human norovirus in gnotobiotic pigs. *J Virol*, *80*, 10372-10381.

Chen, Y. J., Cao, N. X., Xie, R. H., Ding, C. X., Chen, E. F., Zhu, H. P., Sun, J. M., Shang, X. P., Wang, X. X. & Miao, Z. P. (2016). Epidemiological investigation of a tap water-mediated hepatitis E virus genotype 4 outbreak in Zhejiang Province, China. *Epidemiol Infect*, *144*, 3387-3399.

Chen, H. Q., Hoover, D. G. & Kingsley, D. H. (2005). Temperature and treatment time influence high hydrostatic pressure inactivation of feline calicivirus, a norovirus surrogate. *J Food Protect*, *68*, 2389-2394.

Cliver, D. O. & Matsui, S. M. (2002). Viruses. In: Cliver DO, Riemann HP (Eds.), *Foodborne diseases.*, pp. 161-176. Academic Press, Inc., London, UK.

Collier, M. G., Khudyakov, Y. E., Selvage, D., Adams-Cameron, M., Epson, E., Cronquist, A., Jervis, R. H., Lamba, K., Kimura, A. C., Sowadsky, R., Hassan, R., Park, S. Y., Garza, E., Elliott, A. J., Rotstein, D. S., Beal, J., Kuntz, T., Lance, S. E., Dreisch, R., Wise, M. E., Nelson, N. P., Suryaprasad, A., Drobeniuc, J., Holmberg, S. D. & Xu, F. (2014). Outbreak of hepatitis A in the USA associated with frozen pomegranate arils imported from Turkey: an epidemiological case study. *Lancet Infect Dis*, *10*, 976-981.

Cook, N., Bridger, J., Kendall, K., Gomara, M. I., El-Attar, L. & Gray, J. (2004). The zoonotic potential of rotavirus. *J Infect*, *48*, 289-302.

Cromeans, T., Sobsey, M. & Fields, H. (1987). Development of a plaque-assay for a cytopathic, rapidly replicating isolate of hepatitis-A virus. *J Med Virol*, *22*, 45-56.

Cruz-Romero, M., Smiddy, M., Hill, C., Kerry, J. P. & Kelly, A. L. (2004). Effects of high pressure treatment on physicochemical characteristics of fresh oysters (*Crassostrea gigas*). *Innov Food Sci Emerg Technol*, *5*, 161-169.

Dancho, B. A., Chen, H. & Kingsley, D. H. (2012). Discrimination between infectious and non-infectious human norovirus using porcine gastric mucin. *Int J Food Microbiol*, *155*, 222-226.

Desenclos, J. C., Klontz, K. C., Wilder, M. H., Nainan, O. V., Margolis, H. S. & Gunn, R. A. (1991). A multistate outbreak of hepatitis A caused by the consumption of raw oysters. *Am J Public Health*, *10*, 1268-1272.

DiMartino, B., DiProfio, F., Melegari, I., Sarchese, V., Robetto, S., Marsilio, F. & Martella, V. (2016). Detection of hepatitis E virus (HEV) in goats. *Virus Res*, *225*, 69-72.

Donnan, E. J., Fielding, J. E., Gregory, J. E., Lalor, K., Rowe, S., Goldsmith, P., Antoniou, M., Fullerton, K. E., Knope, K., Copland, J. G., Bowden, D. S., Tracy, S. L., Hogg, G. G., Tan, A., Adamopoulos, J., Gaston, J. & Vally, H. (2012). A multistate outbreak of hepatitis A associated with semidried tomatoes in Australia, 2009. *Clin Infect Dis*, *54*, 775-781.

Dumay, E., Picart, L., Regnault, S. & Thiebaud, M. (2006). High pressure–low temperature processing of food proteins. *Biochim Biophys Acta Proteins Proteom*, *1764*, 599-618.

Ettayebi, K., Crawford, S. E., Murakami, K., Broughman, J. R., Karandikar, U., Tenge, V. R., Neill, F. H., Blutt, S. E., Zeng, X. L., Qu, L., Kou, B., Opekun, A. R., Burrin, D., Graham, D. Y., Ramani, S., Atmar, R. L. & Estes, M. K. (2016). Replication of human noroviruses in stem cell–derived human enteroids. *Science*, *353*, 1387-1393.

Farkas, T., Cross, R. W., Hargitt, E., Lerche, N. W., Morrow, A. L. & Sestak, K. (2010). Genetic diversity and histo-blood group antigen interactions of rhesus enteric caliciviruses. *J Virol*, *84*, 8617-8625.

Farkas, T., Sestak, K., Wei, C. & Jiang, X. (2008). Characterization of a rhesus monkey calicivirus representing a new genus of *Caliciviridae*. *J Virol*, *82*, 5408-5416.

Fitzsimons, D., Hendrickx, G., Vorsters, A. & Van Damme, P. (2010). Hepatitis A and E: update on prevention and epidemiology. *Vaccine*, *28*, 583-588.

Fong, T. T. & Lipp, E. K. (2005). Enteric viruses of humans and animals in aquatic environments: Health risks, detection, and potential water quality assessment tools. *Microbiol Mol Biol Rev*, *69*, 357-371.

Gao, S., Li, D., Zha, E., Zhou, T., Wang, S. & Yue, X. (2015). Surveillance of hepatitis E virus contamination in shellfish in China. *Int J Res Public Health*, *12*, 2026-2036.

Giddings, N., Ailard, H. & Hite, B. (1929). Inactivation of the tobacco-mosaic virus by high pressures. *Phytopathology*, *19*, 749-750.

Grodzki, M., Schaeffer, J., Piquet, J. C., Le Saux, J. C., Chevé, J., Ollivier, J., Le Pendu, J. & Le Guyader, F. S. (2014). Bioaccumulation efficiency, tissue distribution, and environmental occurrence of hepatitis E virus in bivalve shellfish from France. *Appl Environ Microbiol*, *80*, 4269-4276.

Gogal, R. M., Jr. Kerr, R., Kingsley, D. H., Granata, L. A., LeRoith, T., Holliman, S. D., Dancho, B. A. & Flick, G. J. (2011). High hydrostatic pressure processing of murine norovirus 1-contaminated oysters inhibits oral infection in STAT-1(-/-)-deficient female mice. *J Food Protect*, *74*, 209-214.

Green, K. Y., Ando, T., Balayan, M. S., Berke, T., Clarke, I. N., Estes, M. K., Matson, D. O., Nakata, S., Neill, J. D., Studdert, M. J. & Thiel, H. J. (2000). Taxonomy of the caliciviruses. *J Infect Dis*, *181*, S322-S330.

Gross, M. & Jaenicke, R. (1994). Proteins under pressure - the influence of high hydrostatic-pressure on structure, function and assembly of proteins and protein complexes. *Eur J Biochem*, *221*, 617-630.

Grove, S. F., Forsyth, S., Wan, J., Coventry, J., Cole, M., Stewart, C. M., Lewis, T., Ross, T. & Lee, A. (2008). Inactivation of hepatitis A virus, poliovirus and a norovirus surrogate by high pressure processing. *Innov Food Sci Emerg Technol*, *9*, 206-210.

Grove, S. F., Lee, A., Lewis, T., Stewart, C. M., Chen, H. Q. & Hoover, D. G. (2006). Inactivation of foodborne viruses of significance by high pressure and other processes. *J Food Protect*, *69*, 957-968.

Hall, A. J., Eisenbart, V. G., Etingüe, A. L., Gould, L. H., Lopman, B. A. & Parashar, U. D. (2012). Epidemiology of foodborne norovirus outbreaks, United States, 2001–2008. *Emerg Infect Dis*, *18*, 1566-1573.

Halliday, M. L., Kang, L., Zhou, T., Hu, M., Pan, Q., Fu, T., Huang, Y. & Hu, S. (1991). An epidemic of hepatitis A attributable to the ingestion of raw clams in Shanghai, China. *J Infect Dis*, *164*, 852-859.

Hayashi, R. (1992). Utilization of pressure in addition to temperature in food science and technology. In: Balny C, Hayashi R, Heremans K, Masson P. (Eds.), *High pressure and biotechnology*, pp. 185-193. Libby Eurotext Ltd., London, UK.

He, H., Adams, R. M., Farkas, D. F. & Morrissey, M. T. (2002). Use of high-pressure processing for oyster shucking and shelf-life extension. *J Food Sci*, *67*, 640-645.

Hendrickx, M., Ludikhuyze, L., van den Broeck, I. & Weemaes, C. (1998). Effects of high pressure on enzymes related to food quality. *Trends Food Sci Technol*, *9*, 197-203.

Hirneisen, K. A. & Kniel, K. E. (2013). Inactivation of internalized and surface contaminated enteric viruses in green onions. *Int J Food Microbiol*, *166*, 201-206.

Hite, B. H. (1899). The effect of pressure in the preservation of milk: a preliminary report. *West Virginia Agricultural Experiment Station*, *58*, 15-35.

Huang, R., Li, X., Huang, Y. & Chen, H. (2014). Strategies to enhance high pressure inactivation of murine norovirus in strawberry puree and on strawberries. *Int J Food Microbiol*, *185*, 1-6.

Huang, F., Li, Y., Yu, W., Jing, S., Wang, J., Long, F., He, Z., Yang, C., Bi, Y., Cao, W., Liu, C., Hua, X. & Pan, Q. (2016). Excretion of infectious hepatitis E virus into milk in cows imposes high risks of zoonosis. *Hepatology*, *64*, 350-359.

Iizuka, S., Oka, T., Tabara, K., Omura, T., Katayama, K., Takeda, N. & Noda, M. (2010). Detection of sapoviruses and noroviruses in an outbreak of gastroenteritis linked genetically to shellfish. *J Med Virol*, *82*, 1247-1254.

Isbarn, S., Buckow, R., Himmelreich, A., Lehmacher, A. & Heinz, V. (2007). Inactivation of avian influenza virus by heat and high hydrostatic pressure. *J Food Protect*, *70*, 667-673.

Jacobsen, K. & Koopman, J. (2004). Declining hepatitis A seroprevalence: a global review and analysis. *Epidemiol Infect*, *132*, 1005-1022.

Jiang, X., Wang, M., Wang, K. N. & Estes, M. K. (1993). Sequence and genomic organization of Norwalk virus. *Virology*, *195*, 51-61.

Jones, M. K., Grau, K. R., Costantini, V., Kolawole, A. O., de Graaf, M., Freiden, P., Graves, C. L., Koopmans, M., Wallet, S. M., Tibbetts, S. A., Schultz-Cherry, S., Wobus, C. E., Vinjé, J. & Karst, S. M. (2015). Human norovirus culture in B cells. *Proc Nat Acad Sci*, *10*, 1939-1947.

Kapikian, A. Z. (2000). The discovery of the 27-nm Norwalk virus: an historic perspective. *J Infect Dis*, *181*, S295-302.

Kars, S. M., Wobus, C. E., Lay, M., Davidson, J. & Virgin, H. W. (2003). STAT-1-dependent innate immunity to a Norwalk-like virus. *Science*, *299*, 1575-1578.

Khadre, M. & Yousef, A. (2002). Susceptibility of human rotavirus to ozone, high pressure, and pulsed electric field. *J Food Protect*, *65*, 1441-1446.

Kingsley, D. H. (2011). Food-borne Noroviruses. In: Fratamico P, Kathariou S, Liu Y. (Eds.) *Genomes of Food- and Water-borne Pathogens*, pp. 237-245. ASM Press, Washington DC.

Kingsley, D. H. (2013). High pressure processing and its application to the challenge of virus-contaminated foods. *Food Environ Virol*, *5*, 1-12.

Kingsley, D. H. (2014a). High pressure processing of bivalve shellfish and HPP's use as a virus intervention. Special Issue: High Pressure Processing of Foods. *Foods*, *3*, 336-350.

Kingsley, D. H. (2014b). Shellfish (Molluscs and Crustacea): shellfish contamination and spoilage. In: Batt CA, Tortorello ML. (Eds.) *Encyclopedia of Food Microbiology*, vol *3*, pp. 389–396. Elsevier Ltd, Academic Press.

Kingsley, D. H. (2016). Emerging food-borne and agriculture-related viruses. *Microbiol Spectrum*, *4*, 1-15.

Kingsley, D. H., Calci, K. R., Holliman, S., Dancho, B. A. & Flick, G. J. (2009a). High pressure inactivation of HAV within oysters: comparison of shucked oysters with whole-in-shell meats. *Food Environ Virol*, *1*, 137-140.

Kingsley, D. H. & Chen, H. (2008). Aqueous matrix compositions and pH influence feline calicivirus inactivation by high pressure processing. *J Food Protect*, *71*, 1598-1603.

Kingsley, D. H. & Chen, H. (2009b). Influence of pH, salt, and temperature on pressure inactivation of hepatitis A virus. *Int J Food Microbiol*, *130*, 61-64.

Kingsley, D. H., Chen, H. Q. & Hoover, D. G. (2004). Inactivation of selected picornaviruses by high hydrostatic pressure. *Virus Res*, *102*, 221-224.

Kingsley, D. H., Guan, D. S. & Hoover, D. G. (2005). Pressure inactivation of hepatitis A virus in strawberry puree and sliced green onions. *J Food Protect*, *68*, 1748-1751.

Kingsley, D. H., Li, X. & Chen, H. (2014b). Temperature effects for high pressure processing of picornaviruses. *Food Environ Virol*, *6*, 58-61.

Kingsley, D. H., Richards, G. P., Papafragkou, E. & Hoover, D. G. (2002). Inactivation of hepatitis A virus and a calicivirus by high hydrostatic pressure. *J Food Protect*, *65*, 1605-1609.

Kingsley, D. H., Duncan, S. E., Granata, L. A., Salinas-Jones, A., Flick, G. J., Bourne, D. M. & Fernadez-Plotka, V. C. (2015). High pressure processing with hot sauce flavouring enhances sensory quality of raw oysters (*Crassostrea virginica*). *Int J Food Sci Technol*, *50*, 2013-2021.

Kingsley, D. H., Guan, D., Hoover, D. G. & Chen, H. (2006). Inactivation of hepatitis A virus by high-pressure processing: The role of temperature and pressure oscillation. *J Food Protect*, *69*, 2454-2459.

Kingsley, D. H., Holliman, D. R., Calci, K. R., Chen, H. & Flick, G. J. (2007). Inactivation of a norovirus by high-pressure processing. *Appl Environ Microbiol*, *73*, 581-585.

Kingsley, D. H., Kuhn, D. D., Flick, G. J., Oh, J., Lawson, L. S., Meade, G. K. & Gieseck, C. C. (2014a). Desirability of oysters treated by high pressure processing at different temperatures and elevated pressures. *Am J Food Technol*, *9*, 209-216.

Kingsley, D. H., Vincent, E., Meade, G. K., Watson, C. & Fan, X. (2014c). Inactivation of Human Norovirus using Chemical Sanitizers. *Int J Food Microbiol*, *171*, 94-99.

Kitamura, Y. & Itoh, T. (1987). Reaction volume of protonic ionization for buffering agents. Prediction of pressure dependence of pH and pOH. *J Sol Chem*, *16*, 715–725.

Knorr, D. (1999). Novel approaches in food-processing technology: new technologies for preserving foods and modifying function. *Curr Opin Biotechnol*, *10*, 485-491.

Koo, H. L., Ajami, N., Atmar, R. L. & DuPont, H. L. (2010). Noroviruses: The leading cause of gastroenteritis worldwide. *Discov Med*, *10*, 61-70.

Koopmans, M. & Duizer, E. (2004). Foodborne viruses: an emerging problem. *Int J Food Microbiol*, *90*, 23-41.

Kovač, K., Bouwknegt, M., Díez-Valcarce, M., Raspor, P., Hernández, M. & Rodríguez-Lázaro, D. (2012). Evaluation of high hydrostatic pressure effect on human adenovirus using molecular methods and cell culture. *Int J Food Microbiol*, *157*, 368-374.

Kunugi, S. & Tanaka, N. (2002). Cold denaturation of proteins under high pressure. *Biochim Biophys Acta Prot Struct Mol Enzymol*, *1595*, 329-344.

Kupferschmidt, K. (2016). Europe's new hepatitis problem. *Science*, *353*, 862-863.

Kural, A. G. & Chen, H. (2008). Conditions for a 5-log reduction of *Vibrio vulnificus* in oysters through high hydrostatic pressure treatment. *Int J Food Microbiol*, *122*, 180-187.

Kural, A. G., Shearer, A. E. H., Kingsley, D. H. & Chen, H. (2008). Conditions for high pressure inactivation of *Vibrio parahaemolyticus* in oysters. *Int J Food Microbiol*, *127*, 1-5.

Lees, D. (2000). Viruses and bivalve shellfish. *Int J Food Microbiol*, *59*, 81-116.

Le Guyader, F. S., Atmar, R. L. & Le Pendu, J. (2012). Transmission of viruses through shellfish: when specific ligands come into play. *Curr Opin Virol*, *2*, 103-110.

Le Guyader, F. S., Le Saux, J., Ambert-Balay, K., Krol, J., Serais, O., Parnaudeau, S., Giraudon, H., Delmas, G., Pommepuy, M., Pothier, P. & Atmar, R. L. (2008). Aichivirus, norovirus, astrovirus, enterovirus, and rotavirus involved in clinical cases from a French oyster-related gastroenteritis outbreak. *J Clin Microbiol 46*, 4011-4017.

Leon, J. S., Kingsley, D. H., Montes, J. S., Richards, G. P., Lyon, G. M., Abdulhafid, G. M., Seitz, S. R., Fernandez, M. L., Teunis, P. F., Flick, G. J. & Moe, C. L. (2011). Randomized, double-blinded clinical trial for human norovirus inactivation in oysters by high hydrostatic pressure processing. *Appl Environ Microbiol*, *77*, 5476-5482.

L'Homme, S., Marion, O., Abravanel, F., Chapuy-Regaud, S., Kamar, N. & Izopet, J. (2016). Hepatitis E pathogenesis. *Viruses*, *8*, 212.

Li, D., Tang, Q., Wang, J., Wang, Y., Zhao, Q. & Xue, C. (2009). Effects of high-pressure processing on murine norovirus-1 in oysters (*Crassostrea gigas*) *in situ*. *Food Cont*, *20*, 992-996.

Li, X. & Chen, H. (2015). Evaluation of the porcine gastric mucin binding assay for high-pressure-inactivation studies using murine norovirus and Tulane virus. *Appl Environ Microbiol*, *81*, 515-521.

Li, X., Chen, H. & Kingsley, D. H. (2013a). The influence of temperature, pH, and water immersion on the high hydrostatic pressure inactivation of GI.1 and GII.4 human noroviruses. *Int J Food Microbiol*, *167*, 138-143.

Li, X., Ye, M., Neetoo, H., Golovan, S. & Chen, H. (2013b). Pressure inactivation of Tulane virus, a candidate surrogate for human norovirus and its potential application in food industry. *Int J Food Microbiol*, *162*, 37-42.

Lin, Y., Fengling, L., Lianzhu, W., Yuxiu, Z. & Yanhua, J. (2014). Function of VP2 protein in the stability of the secondary structure of virus-like particles of genogroup II norovirus at different pH levels. *J Microbiol*, *52*, 970-975.

López-Caballero, M., Pérez-Mateos, M., Montero, P. & Borderías, A. J. (2000). Oyster preservation by high-pressure treatment. *J Food Protect*, *63*, 196-201.

Lou, F., DiCaprio, E., Li, X., Dai, X., Ma, Y., Hughes, J., Chen, H., Kingsley, D. H. & Li, J. (2016). Variable High-Pressure-Processing Sensitivities for Genogroup II Human Noroviruses. *Appl Environ Microbiol*, *82*, 6037-6045.

Lou, F., Ye, M., Ma, Y., Li, X., Chen, H., Krakowka, S., Hughes, J., Kingsley, D. H. & Li, J. (2015). A gnotobiotic pig model for determining human norovirus inactivation by high-pressure. *Appl Environ Microbiol*, *81*, 6679-6686.

Lou, F., Neetoo, H., Chen, H. & Li, J. (2011). Inactivation of a human norovirus surrogate by high-pressure processing: effectiveness, mechanism, and potential application in the fresh produce industry. *Appl Environ Microbiol*, *77*, 1862-1871.

Loutreul, J., Cazeaux, C., Levert, D., Nicolas, A., Vautier, S., Le Sauvage, A. L., Perelle, S. & Morin, T. (2014). Prevalence of human noroviruses in frozen marketed shellfish, red fruits, and fresh vegetables. *Food Environ Virol*, *6*, 157-68.

Lowther, J. A., Avant, J. M., Gizynski, K., Rangdale, R. E. & Lees, D. N. (2010). Comparison between quantitative real-time reverse transcription PCR results for norovirus in oysters and self-reported gastroenteric illness in restaurant customers. *J Food Protect*, *73*, 305-311.

Makison-Booth, C. (2014). Vomiting Larry: a simulated vomiting system for assessing environmental contamination from projectile vomiting related to norovirus infection. *J Infect Prevent*, *15*, 176–180.

Martin, D. E. & Hall, S. G. (2006). Oyster shucking technologies: past and present. *Int J Food Sci Technol*, *41*, 223-232.

Mena, K. D. & Gerba, C. P. (2009). Waterborne adenovirus. *Rev Environ Contam Toxicol*, *198*, 133-67.

Meng, X. (2011). From barnyard to food table: The omnipresence of hepatitis E virus and risk for zoonotic infection and food safety. *Virus Res*, *161*, 23-30.

Mesquita, J. R., Oliveira, D., Rivadulla, E., Abreu-Silva, J., Varela, M. F., Romalde, J. L. & Nascimento, M. S. (2016). Hepatitis E virus genotype 3 in mussels (*Mytilus galloprovinciallis*), Spain. *Food Microbiol*, *58*, 13-15.

Michel, M. & Autio, K. (2002). Effects of high pressure on protein- and polysaccharide-based structures. In: Hendricks MEG, Knorr D. (Eds.) *Ultra high pressure treatments of foods*, pp. 189-214. Academic/Plenum Publishers, New York, NY.

Müller, A., Collineau, L., Stephan, R., Müller, A. & Stärk, K. D. (2016). Assessment of the risk of foodborne transmission and burden of hepatitis E in Switzerland. *Int J Food Microbiol*, *242*, 107-115.

Murchie, L. W., Cruz-Romero, M., Kerry, J. P., Linton, M., Patterson, M. F., Smiddy, M. & Kelly, A. L. (2005). High pressure processing of shellfish: A review of microbiological and other quality aspects. *Innov Food Sci Emerg Technol*, *6*, 257-270.

Murchie, L. W., Kelly, A. L., Wiley, M., Adair, B. M. & Patterson, M. (2007). Inactivation of a calicivirus and enterovirus in shellfish by high pressure. *Innov Food Sci Emerg Technol*, *8*, 213-217.

Nantel-Fortier, N., Letellier, A., Lachapelle, V., Fravalo, P., L'Homme, Y. & Brassard, J. (2016). Detection and phylogenetic analysis of the hepatitis E virus in a Canadian swine production network. *Food Environ Virol*, *8*, 296-304.

Oishi, I., Yamazaki, K., Kimoto, T., Minekawa, Y., Utagawa, E., Yamazaki, S., Inouye, S., Grohmann, G. S., Monroe, S. S., Stine, S. E., Carcamo, C., Ando, T. & Glass, R. I. (1994). A large outbreak of acute gastroenteritis associated with astrovirus among students and teachers in Osaka, Japan. *J Infect Dis*, *170*, 439-443.

Otero, L., Ramos, A. M., de Elvira, C. & Sanz, P. D. (2007). A model to design high-pressure processes towards a uniform temperature distribution. *J Food Eng*, *78*, 1463–1470.

Pina, S., Puig, M., Lucena, F., Jofre, J. & Girones, R. (1998). Viral pollution in the environment and in shellfish: human adenovirus detection by PCR as an index of human viruses. *Appl Environ Microbiol*, *64*, 3376-3382.

Pontes, L., Cordeiro, Y., Giongo, V., Villas-Boas, M., Barreto, A., Araujo, J. R. & Silva, J. L. (2001). Pressure-induced formation of inactive triple-shelled rotavirus particles is associated with changes in the spike protein VP4. *J Mol Biol*, *307*, 1171-1179.

Prasad, B., Hardy, M., Dokland, T., Bella, J., Rossmann, M. & Estes, M. K. (1999). X-ray crystallographic structure of the Norwalk virus capsid. *Science*, *286*, 287-290.

Prescott, L. M., Harley, J. P. & Klein, D. A. (2002). Microbiology, fifth edition. McGraw-Hill Publishing, Inc., New York. USA.

Provost, K., Dancho, B. A., Ozbay, G., Anderson, R., Richards, G. P. & Kingsley, D. H. (2011). Hemocytes are sites of persistence for enteric viruses within oysters. *Appli Environ Microbiol*, *77*, 8360-8369.

Rasanayagam, V., Balasubramaniam, V. M., Ting, E., Sizer, C. E., Bush, C. & Anderson, C. (2003). Compression heating of selected fatty food materials during high-pressure processing. *J Food Sci*, *68*, 254-259.

Rzezutka, A. & Cook, N. (2004). Survival of human enteric viruses in the environment and food. *FEMS Microbiol Rev*, *28*, 441-453.

Scallan, E., Hoekstra, R. M., Angulo, F. J., Tauxe, R. V., Widdowson, M., Roy, S. L., Jones, J. L. & Griffin, P. M. (2011). Foodborne illness acquired in the United States-major pathogens. *Emerg Infect Dis*, *17*, 7-15.

Shearer, A., Dunne, C., Sikes, A. & Hoover, D. (2000). Bacterial spore inhibition and inactivation in foods by pressure, chemical preservatives, and mild heat. *J Food Protect*, *63*, 1503-1510.

Shieh, Y. C., Khudyakov, Y. E., Xia, G., Ganova-Raeva, L. M., Khambaty, F. M., Woods, J. W., Veazey, J. E., Motes, M. L., Glatzer, M. B., Bialek, S. R. & Fiore, A. E. (2007). Molecular confirmation of oysters as the vector for hepatitis A in a 2005 multistate outbreak. *J Food Protect*, *70*, 145-150.

Shimasaki, N., Kiyohara, T., Totsuka, A., Nojima, K., Okada, Y., Yamaguchi, K., Kajioka, J., Wakita, T. & Yoneyama, T. (2009). Inactivation of hepatitis A virus by heat and high hydrostatic pressure: variation among laboratory strains. *Vox Sanguinis*, *96*, 14-19.

Sido, R. F., Huang, R., Lui, C. & Chen, H. (2016). High hydrostatic pressure inactivation of murine norovirus and human norovirus on green onions and salsa. *Int J Food Microbiol*, *242*, 1-6.

Tang, Q., Li, D., Xu, J., Wang, J., Zhao, Y., Li, Z. & Xue, C. (2010). Mechanism of inactivation of murine norovirus-1 by high pressure processing. *Int J Food Microbiol*, *137*, 186-189.

Terio, V., Tantillo, G., Martella, V., Di Pinto, P., Buonavoglia, C. & Kingsley, D. H. (2010). High pressure inactivation of HAV within mussels. *Food Environ Virol*, *2*, 83-88.

Teunis, P. F. M., Moe, C. L., Liu, P., Miller, S. E., Lindesmith, L., Baric, R. S., Le Pendu, J. & Calderon, R. L. (2008). Norwalk virus: How infectious is it? *J Med Virol*, *80*, 1468-1476.

Tian, P., Engelbrektson, A. & Mandrell, R. (2008). Two-log increase in sensitivity for detection of norovirus in complex samples by concentration with porcine gastric mucin conjugated to magnetic beads. *Appl Environ Microbiol*, *74*, 4271-4276.

Tian, S. M., Ruan, K. C., Qian, J. F., Shao, G. Q. & Balny, C. (2000). Effects of hydrostatic pressure on the structure and biological activity of infectious bursal disease virus. *Eur J Biochem*, *267*, 4486-4494.

Tian, P., Yang, D., Jiang, X., Zhong, W., Cannon, J. L., Burkhardt, W. III., Woods, J. W., Hartman, G., Lindesmith, L., Baric, R. S. & Mandrell, R. (2010). Specificity and kinetics of norovirus binding to magnetic bead-conjugated histo-blood group antigens. *J Appl Microbiol*, *109*, 1753-1762.

Ueki, Y., Shoji, M., Okimura, Y., Miyota, Y., Masago, Y., Oka, T., Katayama, K., Takeda, N., Noda, M., Miura, T., Sano, D. & Omura, T. (2010). Detection of Sapovirus in oysters. *Microbiol Immunol*, *54*, 483-486.

Ueki, Y., Shoji, M., Suto, A., Tanabe, T., Okimura, Y., Kikuchi, Y., Saito, N., Sano, D. & Omura, T. (2007). Persistence of caliciviruses in artificially contaminated oysters during depuration. *Appl Environ Microbiol*, *73*, 5698-5701.

Urbanucc, A., Myrmel, M., Berg, I., von Bonsdorff, C. & Maunula, L. (2009). Potential internalisation of caliciviruses in lettuce. *Int J Food Microbiol*, *135*, 175-178.

Wang, C. Y., Huang, H. W., Hsu, C. P. & Yang, B. B. (2016). Recent advances in food processing using high hydrostatic pressure technology. *Crit Rev Food Sci Nutr*, *56*, 527-540.

Wheeler, C., Vogt, T. M., Armstrong, G. L., Vaughan, G., Weltman, A., Nainan, O. V., Dato, V., Xia, G., Waller, K., Amon, J., Lee, T. M., Highbaugh-Battle, A., Hembree, C., Evenson, S., Ruta, M. A., Williams, I. T., Fiore, A. E. & Bell, B. P. (2005). An outbreak of hepatitis A associated with green onions. *New Engl J Med*, *353*, 890-897.

Widdowson, M. A., Sulka, A., Bulens, S. N., Beard, R. S., Chaves, S. S., Hammond, R., Salehi, E. D. P., Swanson, E., Totaro, J., Woron, R., Mead, P. S., Bresee, J. S., Monroe, S. S. & Glass, R. I. (2005). Norovirus and foodborne disease, United States, 1991-2000. *Emerg Infect Dis*, *11*, 154-158.

Wilkinson, N., Kurdziel, A. S., Langton, S., Needs, E. & Cook, N. (2001). Resistance of poliovirus to inactivation by high hydrostatic pressures. *Innov Food Sci Emerg Technol*, *2*, 95-98.

Wilson, D. R., Dabrowski, L., Stringer, S., Moezelaar, R. & Brocklehurst, T. F. (2008). High pressure in combination with elevated temperature as a method for the sterilisation of food. *Trends Food Sci Technol*, *19*, 289-299.

Wobus, C. E., Thackray, L. B. & Virgin, H. W. (2006). Murine norovirus: a model system to study norovirus biology and pathogenesis. *J Virol*, *80*, 5104-5112.

Wu, H. M., Fornek, M., Schwab, K. J., Chapin, A. R., Gibson, K., Schwab, E., Spencer, C. & Henning, K. (2005). A norovirus outbreak at a long-term-care facility: the role of environmental surface contamination. *Infect Control Hosp Epidemiol*, *26*, 802-810.

Yamashita, T., Kobayashi, S., Sakac, K., Nakata, S., Chiba, S., Ishihara, Y. & Isomura, S. (1991). Isolation of cytopathic small round viruses with BS-C-l cells from patients with gastroenteritis. *J Infect Dis*, *164*, 954-957.

Yamashita, T., Sugiyama, M., Tsuzuki, H., Sakae, K., Suzuki, Y. & Miyazaki, Y. (2000). Application of a reverse transcription-PCR for identification and differentiation of Aichi virus, a new member of the Picornavirus family associated with gastroenteritis in humans. *J Clin Microbiol*, *38*, 2955-2961.

Yamashita, T., Sakae, K., Tsuzuki, H., Suzuki, Y., Ishikawa, N., Takeda, N., Miyamura, T. & Yamazaki, S. (1998). Complete nucleotide sequence and genetic organization of Aichi virus, a distinct member of the *Picornaviridae* associated with acute gastroenteritis in humans. *J Virol*, *72*, 8408-8412.

Ye, M., Huang, Y. & Chen, H. (2012). Inactivation of *Vibrio parahaemolyticus* and *Vibrio vulnificus* in oysters by high-hydrostatic pressure and mild heat. *Food Microbiol*, *32*, 179-184.

Ye, M., Huang, Y., Gurtler, J. B., Niemira, B. A., Sites, J. E. & Chen, H. (2013). Effects of pre- or post-processing storage conditions on high-hydrostatic pressure inactivation of *Vibrio parahaemolyticus* and *V. vulnificus* in oysters. *Int J Food Microbiol*, *163*, 146-152.

Ye, M., Li, X., Kingsley, D. H., Jiang, X. & Chen, H. (2014). Inactivation of human norovirus in contaminated oysters and clams by high hydrostatic pressure. *Appl Environ Microbiol*, *80*, 2248-2253.

Ye, M., Lingham, T., Huang, Y., Ozbay, G., Ji, L., Karwe, M. & Chen, H. (2015). Effects of high-hydrostatic pressure on inactivation of human norovirus and physical and sensory characteristics of oysters. *J Food Sci*, *80*, M1330-M1335.

Yi, J., Sederdahl, B. K., Wahl, K., Jerris, R. R., Kraft, C. S., McCracken, C., Gillespie, S., Kirby, A. E., Shane, A. L., Moe, C. L. & Anderson, E. J. (2016). Rotavirus and norovirus in pediatric healthcare-associated gastroenteritis. *Open Forum Infect Dis*, *3*, 1-3.

In: Noroviruses
Editor: Jesús L. Romalde
ISBN: 978-1-53611-939-8

Chapter 8

APTAMER: THE POTENTIAL APPLICATION TO NOROVIRUS RESEARCH, DIAGNOSIS, AND THERAPEUTICS

Seiya Hirano, Daisuke Sano, Satoshi Okabe and Masaaki Kitajima*
Division of Environmental Engineering, Faculty of Engineering,
Hokkaido University, Sapporo, Hokkaido, Japan

ABSTRACT

Aptamer has been attracting much attention as a new biomolecular tool for viral diagnostic and therapeutic applications. Aptamer is a functional nucleic acid that folds into a unique three-dimensional conformation that binds to a specific target, which can be an alternative for antibodies. Rather, aptamer offers various advantages over antibodies as a target-specific recognition molecule: high affinity, specificity, and thermostability, low production cost, etc. Several studies have reported

* Corresponding Author address: Division of Environmental Engineering, Faculty of Engineering, Hokkaido University, North 13 West 5, Kita-ku, Sapporo, Hokkaido 060-8628, Japan. Email: mkitajima@eng.hokudai.ac.jp.

DNA aptamers for certain human norovirus strains as well as murine norovirus, which has enabled development and evaluation of aptasensor (aptamer-based biosensor) technology for rapid and sensitive norovirus detection. Therapeutic application of aptamer to inhibiting norovirus infections has not been clearly demonstrated to date, but there have been a few reports suggesting the potential of aptamer-based therapeutics. Taken together, it has been implied that aptamer is a new target-specific biomolecular tool that offers various potential applications to norovirus research, diagnosis, and therapeutics, which should contribute to the infection and disease control of norovirus by rapidly identifying potential infection routes and reducing the disease burden through therapeutic approach. Herein we summarize recent development and application of norovirus-specific aptamers and discuss future research directions towards better control of norovirus transmission/infections using this aptamer technology.

Keywords: Aptamer, norovirus, aptasensor, SELEX

INTRODUCTION

Noroviruses, members of the family *Caliciviridae*, are a major cause of nonbacterial acute gastroenteritis in humans worldwide. Noroviruses infect all age groups and are highly stable in the environment (Robilotti, 2015); therefore, human noroviruses cause secondary infections and large outbreaks by transmitting through contaminated food and water and spreading via person-to-person transmission (Patel et al., 2009). Norovirus outbreaks are often reported to occur in closed environments such as hospitals, nursing homes, cruise ships, and the military bases (Robilotti, 2015). It is estimated that noroviruses cause 699 million illnesses and 219 thousand deaths across all ages per year in the world, resulting in serious ecomomic losses ($4.2 billion direct health system costs and $56.2 billion in productivity losses) (Bartsch et al., 2016).

Several characteristics of noroviruses make it difficult to control and inflict a loss on human society: (i) low infectious dose of noroviruses ($<10^2$ viral particles (Patel et al., 2009)); (ii) prolonged duration of viral shedding, even after symptoms resolve, which increases the risk of secondary infections; (iii) stability of the virus in relatively high concentrations of chlorine (Duizer et al., 2004) and at a wide range of temperatures (from freezing to 60°C) (Cannon et al., 2006); and (iv) high diversity of noroviruses - strains belonging to different genogroups have only 51–56% genomic nucleotide similarity (Katayama et al., 2002). Therefore, diagnostic tools of noroviruses should be specific yet broadly reactive, sensitive, and fast.

Capability of rapid and sensitive norovirus detection is essential to prevent norovirus infections. Due largely to lack of a routine *in vitro* cultivation system for human noroviruses (although successful replication of human noroviruses in cell culture was reported recently (Ettayebi et al., 2016; Jones et al., 2014)), molecular biological techniques, such as PCR-based methods, have widely been used for clinical diagnosis and environmental detection of noroviruses (Patel et al., 2009). However, PCR-based methods still require process time (at least a few hours), laboratory equipment, and skilled personnel; therefore PCR offers limited on-site applications (Vinjé, 2015). Enzyme immunoassay (EIA) is less sensitive than PCR but is more suitable for rapid parallel screening (60-90 min) of the presence of norovirus antigens in many stool samples collected from outbreak(s). Immunochromatographic (ICG) assays enable more rapid (15 min) and bedside diagnosis without laboratory equipment and skilled personnel (Vinjé, 2015), but it is even less sensitive than EIA. There is still a strong need for a specific, rapid, and sensitive norovirus detection/diagnosis method.

Recently, aptamer has been attracting much attention as a new biomolecular tool, which could be used as highly sensitive and rapid diagnostic methods as well as therapeutic tools. In this chapter, we summarize recent development and application of norovirus-specific aptamers and discuss future research directions towards better control of norovirus transmission/infections using aptamer technology.

APTAMER

Aptamers are single-stranded folded nucleic acids (DNA or RNA) able to specifically bind to a target molecule with high affinity, which was first reported in 1990 (Ellington and Szostak, 1990; Tuerk and Gold, 1990). Aptamers with a specific function can be identified and artificially synthesized *in vitro* as detailed in the following section. Aptamers also exist in the natural world known as riboswitch, which is an RNA-based genetic regulation mechanism in bacteria reported in 2002 (Winkler et al., 2002). The term "aptamer" is derived from the Latin word "aptus" meaning to fit (Ellington and Szostak, 1990). As the name implies, aptamers fold into a unique three-dimensional conformation complementary to their targets, such as proteins, peptides, organic dyes, amino acids, drugs, antibiotics, heavy metals, bacterial cells, and virus particles (Jayasena, 1999; Song et al., 2008). Aptamers can form a variety of complex, thermodynamically stable structures mainly geverned by Watson and Crick base-pairing (e.g., hairpin, G-quartet/quadruplex, stem-bulge, pseudoknot, T-junction, and combination of these), which plays an essential role in specific binding with a ligand (de los Santos Alvarez et al., 2008). The specific target-aptamer interactions are complicated because of its highly complex structure, but it is generally understood that aptamer recognizes a target molecule by multiple interactions, such as stacking, shape complementarity, hydrogen bonds, and hydrophobic and electrostatic interactions (Hermann and Patel, 2000).

Aptamers have been attracting much attention as a new biomolecule for viral diagnostics and therapeutic applications, which can be an alternative for antibodies. Antibody-based diagnostic assays have been used for viral antigen detection since 1950s (Yalow and Berson, 1959) and became more widely used after the discovery of a monoclonal antibody and establishment of procedure for mass production of pure antibodies in 1970s (Köhler and Milstein, 1975). Antibodies have been the most widely used recognition molecule for more than four decades, but there are still limitations with antibodies: (i) antibody identification and production processes are time-consuming, labor-intensive, and expensive; (ii) antibodies against substances with high toxicity or without immunogenicity cannot be produced as the identification and production processes takes place *in vivo*; (iii) quality of antibodies tend to vary between different batches; (iv) antibodies are sensitive to temperature and has limited shelf/storage life (Jayasena, 1999). In contrast, aptamer offers various advantages over antibodies as a target-specific recognition molecule:

(i) higher affinity with a ligand - for example, an aptamer for vascular endothelial growth factor showed higher affinity than monoclonal antibody for the same target (Lee et al., 2005; Wu et al., 2006);
(ii) inexpensive and large-scale *in vitro* production is possible by PCR or asymmetric PCR (Bruno, 2015);
(iii) higher flexibility against various targets, including fast-evolving viral agents such as noroviruses;
(iv) little batch-to-batch variations because of highly reproducible *in vitro* synthetic process (Jayasena, 1999);
(v) higher stability under a wide range of temperature with capability of reversible heat denaturation, which potentially enables repeated target binding and release cycles;
(vi) ease of chemical modification with fluorescent dye, linker molecules, etc. (Jayasena, 1999).

Target-Specific Aptamer Selection

Aptamer for a specific target is identified by a process called SELEX (Systematic Evolution of Ligands by EXponential enrichment), which is an *in vitro* selection process whose idea is similar to the natural selection and evolution process discovered by Darwin. Specifically, the SELEX process identifies aptamer as a survivor of repeating cycle consisting of chemical synthesis of nucleotide library, incubation with target, exclusion of unbound sequence, elution of bound sequence, and amplification by (reverse transcription-)PCR (Figure 1).

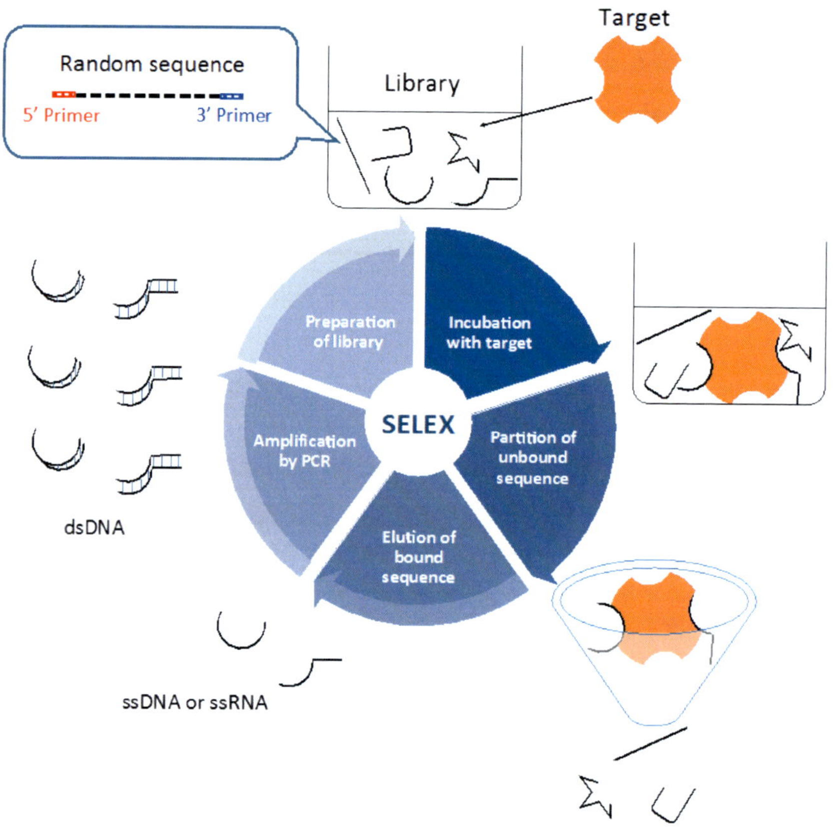

Figure 1. SELEX process.

Figure 1 shows the SELEX process, which begins from artificially synthesized random oligonucleotide library. Theoretically the library can contain up to 4^n different nucleotide sequences (where n is the length (i.e., number of nucleotide) of the oligonucleotide), but the number of molecules is usually limited to the order of 10^{16} in a typical laboratory setting experiment (Jijakli et al., 2016). These nucleotides have an internal randomized sequence flanked by fixed sequences, which serve as primer binding sites in the PCR amplification step (Stoltenburg et al., 2007).

Next, these nucleotides are incubated with the target molecules (positive-selection) or non-target molecules (counter-selection). For virus-specific aptamer, this step is challenging owing to difficulty of purification of virus as a target (van den Kieboom et al., 2015). In most cases, counter-selection is carried out to increase the specificity by excluding aptamers binding to a non-target ligand (e.g., molecules with similar structures to the target). After incubation with a target, separation of bound sequences and unbound sequences is carried out with affinity column, microtiter plates, filter binding, magnetic beads, or biacore. Thereafter, bound sequences are eluted by changing temperature or ion salt concentration or using chaotropic agent (Mairal et al., 2008).

The eluted nucleic acids ("survivors") are amplified by (reverse transcription-)PCR. The enriched dsDNA strands are used to produce the single-stranded oligonucleotide library for the next SELEX round. Usually 8 to 15 rounds are necessary for selection of a target-specific aptamer (Jayasena, 1999). These nucleic acids are sequenced by Sanger sequencing method or next-generation sequencing for determination of the enriched sequences and subsequent mass production via chemical synthesis. Aptamers can be identified in days by using an automated robotic SELEX plaform (Cox et al., 1998).

Norovirus-Specific Aptamer

Several norovirus-specific aptamers targeting structural proteins of norovirus particles have been reported recently (see Table 1). An aptamer for murine norovirus (MNV), the most widely used surrogate/model for human norovirus, was reported in 2013 (Giamberardino et al., 2013). Positive-selection using MNV was carried out for 7 rounds; in order to obtain a strongly binding pool, the stringency of the selections was increased by decreasing target amount (i.e., 500, 100, and 10 infective units were used for SELEX round 1~3, 4~5, and 6~7, respectively). The retained sequences were then subjected to counter-selections with feline calicivirus and culturing medium (Dulbecco's Modified Eagle Medium

(DMEM)) to eliminate any binders with affinity for related targets. One ssDNA sequence (named AG3) selected from potential aptamer candidates showed high binding affinity for MNV.

A DNA aptamer against the norovirus GII.4 major capsid protein (viral protein 1, VP1) was reported in 2014 (Beier et al., 2014). This aptamer was selected based on a modified SELEX method, which developed sequence library in each SELEX round by computational analysis. In this SELEX, two highly dominant sequences in the end of SELEX were already found in round 4. From this result, this method offers a possibility to reduce the required number of rounds and time for SELEX.

Aptamers for human norovirus GII.2 were also reported in 2014 (Escudero-Abarca et al., 2014). The sequences that could bind to whole norovirus GII.2 particles were selected by magnetic beads, and eluted by heating. After finishing each round of SELEX, counter-SELEX was carried out against negative stool specimens or bead-antibody complex. As a result, 34 unique aptamer candidate sequences were identified at the end of the SELEX and counter-SELEX round 4, 7 and 9. Among the 34 candidates, four candidates were selected for further characterization because these candidates were more abundant in the aptamer candidate pool and showed stronger affinities for virus-like particles (VLPs) of human norovirus than the other candidates.

Aptamers for norovirus GII.4 P-domain were developed in 2015 by similar SELEX to the process to develop the above-mentioned aptamers for GII.2 (Moore et al., 2015). In this study, 7 rounds of positive-selection SELEX were performed, followed by 2 rounds of counter-SELEX. DNA pools were sequenced and 11 sequences were identified including 6 types of aptamers at the end of SELEX and counter-SELEX. Aptamers named M1 and M6-2 were selected for a comprehensive set of reasons (i.e., abundance of sequence in the pools, low ΔG values, same motifs in secondary structure). More recently, the potential of the M6-2 aptamer for discriminating norovirus particle functionality was demonstrated (Moore et al., 2016). Specifically, the M6-2 aptamer recognized heating damage of VLP without statistically significant difference from histo-blood group antigens (HBGAs).

Table 1. Reported norovirus-specific aptamers

Name	Sequence (5'–3')	SELEX ligand[a]	Reference
AG3	GCTAGCGAATTCCGTACGAAGGGCGAATTCCACATTGGGCTGCAGCCCGGGGGATCC	MNV	Giamberardino et al., 2013
M1	AGTATACGTATTACCTGCAGCTGTTTATGGGGATAAACGTATCTAATTCGTGTACTAA-TCACGATATCTCGGAGATCTTGC	GII.4 P domain	Moore et al., 2015
M6-2	AGTATACGTATTACCTGCAGCTGGGAAGAGGTCCGGTAAATGCAGGGTCAGCCCGGA-GAGCGATATCTCGGAGATCTTGC	GII.4 P domain	Moore et al., 2015
SMV-21	AGTATACGTATTACCTGCAGCCCATGTTTTGTAGGTGTAATAGGTCATGTTAGGGTTT-CTGCGATATCTCGGAGATCTTGC	SMV (GII.2) VLP	Escudero-Abarca et al., 2014
SMV-25	AGTATACGTATTACCTGCAGCCATCTGTGTGAAGACTATATGGCGCTCACATATTTCT-TTCCGATATCTCGGAGATCTTGC	SMV (GII.2) VLP	Escudero-Abarca et al., 2014
No name	GTCTG TAGTA GGGAG GATGG TCCGG GGCCC CGAGA CGACG TTATC AGGC	GII.4 VLP	Beier et al.,2014

[a] SMV, Snow Mountain virus; VLP, virus-like particle.

Table 2. Dissociation constant values of norovirus aptamers in comparison with other recognition molecules

Name	Ligand VLP[a]	K_d (nM)	Method[b]	Reference
Norovirus aptamers (SMV-19, 21, 25, 26)	GII.2	101-281	ELASA	Escudero-Abarca et al., 2014
Monoclonal antibodies (NV23, NV37, NS22, 3901, 3912, 8812)	GI.1	0.86-19.6	SPR	Kou et al., 2015
Monoclonal antibodies (NV23, NV57, NS14, NS22, NS46)	GII.2	0.30-45	SPR	Kou et al., 2015
Monoclonal antibodies (NV23, NV57, NS14, NS22, NS47)	GII.4	0.49-27	SPR	Kou et al., 2015
Histo-blood group antigen (HBGA) (B trisaccharide, A trisaccharide, B type 6 tetrasaccharide, A type 6 tetrasaccharide)	GII.4	250-1000	ESI-MS	Han et al., 2015
Bacterial HBGA-like substances	GII.6	323	ELISA	Miura et al., 2013
Enteric virus-binding protein	GII.4	210	QCM	Imai et al., 2011

[a] VLP, virus-like particle.

[b] ELASA, enzyme-linked aptamer sorbent assay; ELISA, enzyme-linked immunosorbent assay; SPR, surface plasmon resonance; ESI-MS, electrospray ionization mass spectrometry; QCM, quartz crystal microbalance.

Table 3. Relative affinity of the reported norovirus-specific aptamers for norovirus-like particles

Name	Relative affinity for VLP[a]														Reference
	GI.1	GI.4	GI.6	GI.7	GI.8	GII.1	GII.2	GII.3	GII.4(G)[b]	GII.4(H)[c]	GII.6	GII.7	GII.12	GII.17	
M1	3.4	3.17	1.98	**7.32**	2.61	3.59	**10.68**	2.94	**7.59**	**10.41**	2.38	**7.47**	4.47	3.94	Moore et al., 2015
M6-2	4.28	3.36	2.75	**7.56**	4.55	4.52	**12**	3.16	**11.54**	**12.98**	4.4	**8.02**	**5.55**	**5.24**	Moore et al., 2015
SMV-21	2.4	4.3	3.8	**8.1**	4.6	**8.6**	**18.1**	**11.8**	**6.4**	**11.3**	3	**13.5**	**6.6**	**12.2**	Escudero-Abarca et al., 2014
SMV-25	3	**5.6**	2.4	3.5	**6.2**	**9.4**	**12.4**	**5.4**	**10.4**	**11**	**7.3**	**9.1**	1.9	2	Escudero-Abarca et al., 2014

[a] Values indicate the ratio of abosrbance readings for the sample with virus-like particles (VLPs) versus negative control (without VLPs) obtained from ELASA; Bolded values corresponding to >5.0 indicate medium to strong binding accoding to the literature (Escudero-Abarca et al., 2014; Moore et al., 2015).

[b] GII.4(G), GII.4 Grimsby.

[c] GII.4(H), GII.4 Houston.

AG3 showed high affinity for MNV in low picomolar range dissociation constant as determined by fluorescence anisotropy. This affinity seems so high compared with other molecules binding to norovirus (see Table 2); therefore, this dissociation constant can be considered a multivalent K_d. The AG3 aptamer also showed a million folds higher binding affinity for MNV than the affinity of non-specific DNA for MNV. Interestingly, AG3 displayed considerable binding affinity for a GII.3 human norovirus strain as well (Giamberardino et al., 2013). The binding affinities of the SMV and GII.4 aptamers were determined using VLPs of a number of human norovirus strains belonging to GI.1, GI.4, GI.6, GI.7, GI.8, GII.1, GII.2, GII.3, GII.4, GII.6, GII.7, GII.12, and GII.17. The affinities were evaluated by enzyme-linked aptamer sorbent assay (ELASA), whose working principle is similar to the enzyme-linked immunosorbent assay (ELISA) but ELASA uses aptamers as opposed to antibodies (see Table 3). A candidate aptamer named SMV-21 had affinities for GI.7, GII.1, GII.2, GII.3, GII.4, GII.7, GII.12, and GII.17 and another candidate aptamer named SMV-25 had affinities for GI.4, GI.8, GII.1, GII.2, GII.3, GII.4, GII.6, and GII.7. Two sequences of candidates SMV-19 and SMV-26 exhibited less binding affinity for diverse strains (Escudero-Abarca et al., 2014). Both candidates of GII.4 aptamers M1 and M6-2 showed strong binding affinities for VLPs of human norovirus strains GI.7, GII.2, two GII.4, and GII.7 as determined by ELASA (Moore et al., 2015).

Biosensor is a potential device for a specific, rapid, and sensitive detection/diagnosis method. An MNV aptamer AG3 was immobilized on gold electrodes and used as a recognition molecule of an electrochemical sensor for MNV detection. The result demonstrated that this aptasensor could detect as low as 180 virus particles, demonstrating high detection sensitivity (Giamberardino et al., 2013). Microelectromechanical systems (MEMS)-based devices for detecting MNV using the aptamer AG3 in a solution by electrochemical analysis were developed in 2016 (Kitajima et al., 2016).

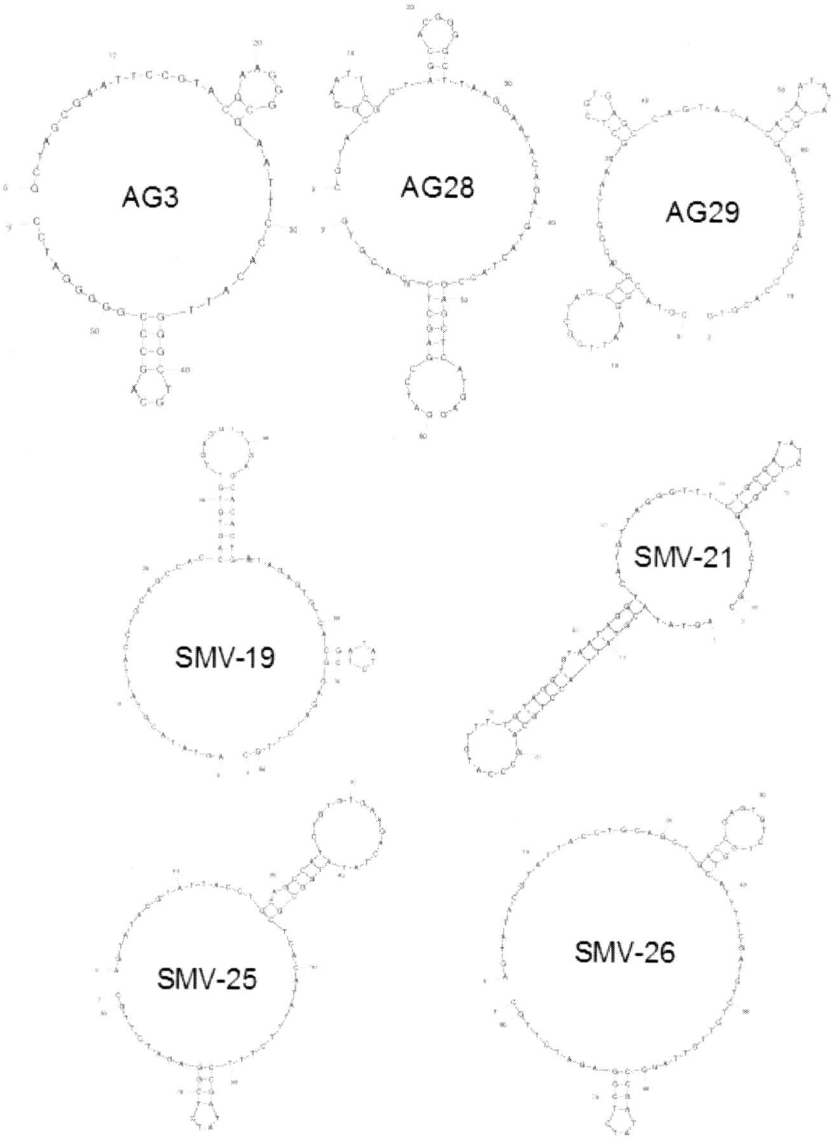

Figure 2 (A). Predicted secondary structures of the reported norovirus-specific aptamers (Escudero-Abarca et al., 2014; Giamberardino et al., 2013) at 37°C in the presence of 50 mM Na^+. The structures were predicted with Mfold (http://mfold.rna.albany.edu/).

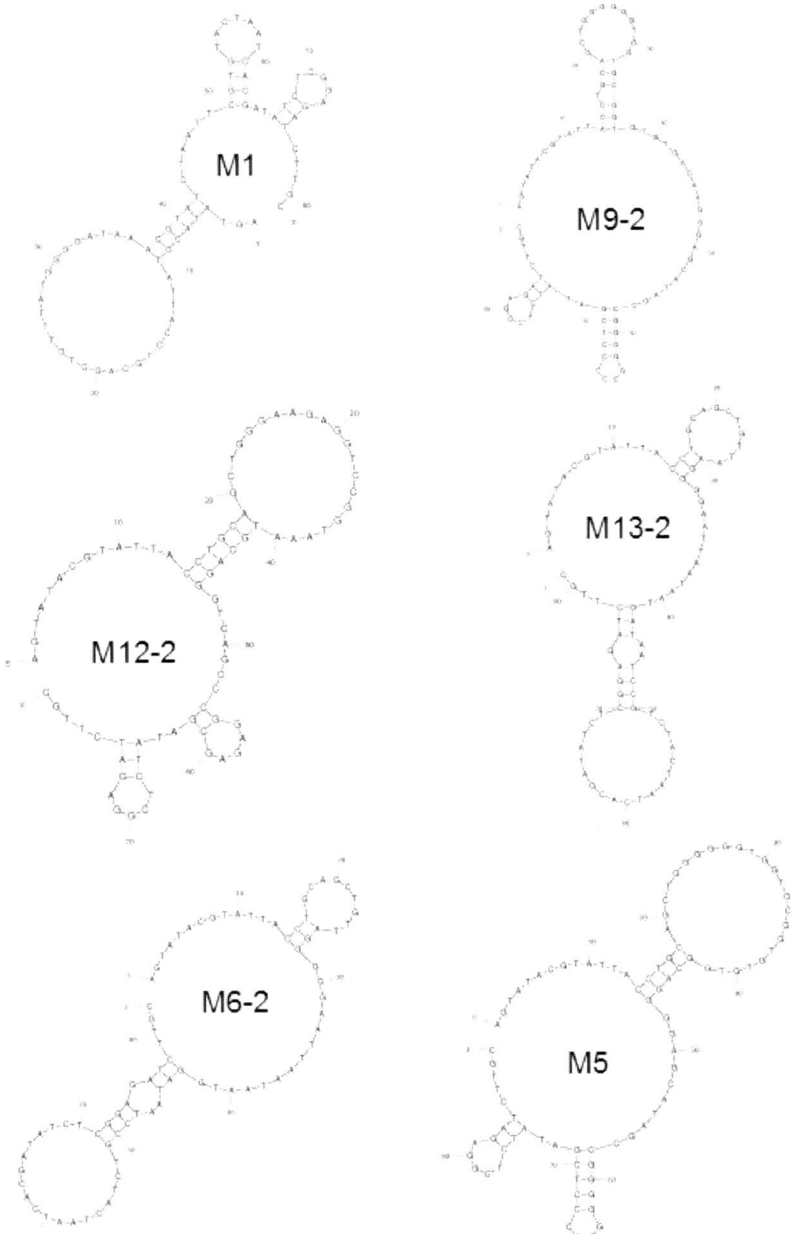

Figure 2 (B). Predicted secondary structures of the reported norovirus-specific aptamers (Moore et al., 2015) at 23°C in the presence of 1.0 mM Na^+ and 0.5 mM Mg^{2+}. The structures were predicted with Mfold (http://mfold.rna.albany.edu/).

The magnitude current difference between peak and baseline decreased linearly with increasing concentrations of MNV by using this electrochemical aptasensor platform (Kitajima et al., 2016). This aptasensor could be a portable device owing to its potential for further simplification and miniaturization.

APTAMER-BASED THERAPEUTICS

Antiviral therapeutics to counter norovirus diseases could be useful for treatment of chronic infections in immunocompromised patients, long-lasting outbreaks in closed settings (hospitals, nursing homes, rafting trips, cruise ships, etc.), and acute illness especially among infants, elderly, and hospitalized patients. Currently there are no effective antiviral therapies available for norovirus infections, but in principle, both viral and host factors and any step in the viral replication cycle can be targeted for the development of norovirus antivirals (Prasad et al., 2016). It was reported that antibodies inhibiting attachment of Norwalk virus particles to host-cell HBGA by blocking epitopes of P subdomain successfully prevented infection *in vivo* in a chimpanzee model (Chen et al., 2013). Aptamer is also attracting attention as one of the candidate molecules to counter viral infections by (i) interfering the interaction between the virus and the host using aptamers targeting receptors on the host-cell surface or viral proteins that bind to the receptor, or (ii) interfering with the virus replication intracellularly, targeting proteins implicated in the viral replication cycle (González et al., 2016; Shum et al., 2013). In fact, therapeutic applications of aptamers to human immunodeficiency virus, hepatitis C virus (HCV), influenza viruses, etc. have been extensively studied recently (Shum et al., 2013). There are many additional advantages of using aptamer over antibodies in therapy: for example, *in vitro* chemical production process is not prone to viral or bacterial contamination, and smaller size allows more efficient delivery to inside biological compartments across membranes (Keefe et al., 2010).

Therapeutic application of aptamer to inhibiting norovirus infections has not been clearly demonstrated to date, but there have been a few reports implying the potential of aptamer-based therapeutics. Jones et al., (2006) reported that aptamers designed to inhibit HCV RNA-dependent RNA polymerase (RdRp), which plays an essential role in synthesizing both anti-sense (negative-strand) RNA and newly-made genomic and subgenomic RNA, showed some inhibitory effect on RdRp activity of norovirus as well (Jones et al., 2006). This result highlights the possibility of developing an aptamer that prevents norovirus infections by inhibiting norovirus RdRp activity *in vivo*. Another possible approach to inhibit norovirus infection would be to block the attachment and internalization of virion into the host cell by using an aptamer binding to protruding parts of capsid protein by forming clefts. Aptamers targeting the protruding P2 subdomain have been developed (Beier et al., 2014a, 2014b), which could be used to control the norovirus infection extracellularly by inhibiting attachment to host cell receptors and/or intracellularly by blocking capsid rearrangements necessary for virion production (Beier et al., 2014b).

FUTURE PERSPECTIVES

Aptamer technology was invented in the early 90s and has undergone major advancements in the last decade. However, only a limited number of studies have paid attention to potential application of aptamers in the field of norovirus research, diagnosis, and therapeutics, probably due to the ignorance that still exists towards this technology. As reviewed in this chapter, aptamer technology offer vast potential to control norovirus infections and mitigate burdens to human society. For instance, application of aptamers to rapid and sensitive norovirus disgnosis/sensing is a promising area of research and the developed aptasensor should contribute significantly to controlling risk of norovirus infections. We may also be able to counter outbreaks due to a new norovirus strain by quickly designing aptamers against the new strain and utilizing the aptamer for diagnosis and antiviral treatment. Future research should focus on the development of aptasensor platforms towards real-world application (clinical diagnosis, food safety, and environmental monitoring), more in-depth understanding of the ability of aptamer to evaluate the capsid functionality and infectivity of human noroviruses, and applicability of aptamers as an antiviral to inhibit norovirus infections *in vivo*.

REFERENCES

Bartsch SM, Lopman BA, Ozawa S, Hall AJ, Lee BY. 2016. Global economic burden of norovirus gastroenteritis. *PLoS One* 11: e0151219. doi:10.1371/journal.pone.0151219.

Beier R, Boschke E, Labudde D. 2014a. New strategies for evaluation and analysis of SELEX experiments. *Biomed Res Int* 2014: 849743. doi:10.1155/2014/849743.

Beier R, Pahlke C, Quenzel P, Henseleit A, Boschke E, Cuniberti G, Labudde D. 2014b. Selection of a DNA aptamer against norovirus capsid protein VP1. *FEMS Microbiol Lett* 351: 162–169. doi:10.1111/1574-6968.12366.

Bruno JG. 2015. Predicting the uncertain future of aptamer-based diagnostics and therapeutics. *Molecules* 20: 6866–6887. doi:10.3390/molecules20046866.

Cannon JL, Papafragkou E, Park GW, Osborne J, Jaykus LA, Vinjé J. 2006. Surrogates for the study of norovirus stability and inactivation in the environment: A comparison of murine norovirus and feline calicivirus. *J Food Prot* 69: 2761–2765.

Chen Z, Sosnovtsev SV, Bok K, Parra GI, Makiya M, Agulto L, Green KY, Purcell RH. 2013. Development of Norwalk virus-specific monoclonal antibodies with therapeutic potential for the treatment of Norwalk virus gastroenteritis. *J Virol* 87: 9547–57. doi:10.1128/JVI.01376-13.

Cox JC, Rudolph P, Ellington A. 1998. Automated RNA selection. *Biotechnol Prog* 14: 845–850. doi:10.1021/bp980097h.

de-los-Santos-Alvarez N, Lobo-Castanon MJ, Miranda-Ordieres AJ, Tunon-Blanco P. 2008. Aptamers as recognition elements for label-free analytical devices. *Trends Anal Chem* 2: 437–446. doi:10.1016/j.trac.2008.03.003.

Duizer E, Bijkerk P, Rockx B, De Groot A, Twisk F, Koopmans M. 2004. Inactivation of caliciviruses. *Appl Environ Microbiol* 70: 4538–4543. doi:10.1128/AEM.70.8.4538.

Ellington AD, Szostak JW. 1990. *In vitro* selection of RNA molecules that bind specific ligands. *Nature* 346: 818–822. doi:10.1038/346818a0.

Escudero-Abarca BI, Suh SH, Moore MD, Dwivedi HP, Jaykus, LA. 2014. Selection, characterization and application of nucleic acid aptamers for the capture and detection of human norovirus strains. *PLoS One* 9: e106805. doi:10.1371/journal.pone.0106805.

Ettayebi K, Crawford SE, Murakami K, Broughman JR, Karandikar U, Tenge VR, Neill FH, Blutt SE, Zeng XL, Qu L, Kou B, Opekun AR, Burrin D, Graham DY, Ramani S, Atmar RL, Estes MK. 2016. Replication of human noroviruses in stem cell-derived human enteroids. *Science* 30: 25–33. doi:10.1126/science.aaf5211.

Giamberardino A, Labib M, Hassan EM, Tetro JA, Springthorpe S, Sattar SA, Berezovski MV, DeRosa MC. 2013. Ultrasensitive norovirus detection using DNA aptasensor technology. *PLoS One* 8: e79087. doi:10.1371/journal.pone.0079087.

González V, Martín M, Fernández G, García-Sacristán A. 2016. Use of aptamers as diagnostics tools and antiviral agents for human viruses. *Pharmaceuticals* 9:78. doi:10.3390/ph9040078.

Han L, Kitova EN, Tan M, Jiang X, Pluvinage B, Boraston AB, Klassen JS. 2015. Affinities of human histo-blood group antigens for norovirus capsid protein complexes. *Glycobiology* 25: 170-180. doi:10.1093/glycob/cwu100.

Hermann T, Patel DJ. 2000. Adaptive recognition by nucleic acid aptamers. *Science* 287: 820–825. doi:10.1126/science.287.5454.820.

Imai T, Sano D, Miura T, Okabe S, Wada K, Masago Y, Omura T. 2011. Adsorption characteristics of an enteric virus-binding protein to norovirus, rotavirus and poliovirus. *BMC Biotechnol* 11: 123. doi:10.1186/1472-6750-11-123.

Jayasena SD. 1999. Aptamers: An emerging class of molecules that rival antibodies in diagnostics. *Clin Chem* 45: 1628–1650.

Jijakli K, Khraiwesh B, Fu W, Luo L, Alzahmi A, Koussa J, Chaiboonchoe A, Kirmizialtin S, Yen L, Salehi-Ashtiani K. 2016. The *in vitro* selection world. *Methods* 106: 3–13. doi:10.1016/j.ymeth.2016.06.003.

Jones LA, Clancy LE, Rawlinson WD, White PA. 2006. High-affinity aptamers to subtype 3a hepatitis C virus polymerase display genotypic specificity. *Antimicrob Agents Chemother* 50: 3019–3027. doi:10.1128/AAC.01603-05.

Jones MK, Watanabe M, Zhu S, Graves CL, Keyes LR, Grau KR, Gonzalez-Hernandez MB, Iovine NM, Wobus CE, Vinjé J, Tibbetts SA, Wallet SM, Karst SM. 2014. Enteric bacteria promote human and mouse norovirus infection of B cells. *Science* 346: 755-759. doi:10.1126/science.1257147.

Katayama K, Shirato-Horikoshi H, Kojima S, Kageyama T, Oka T, Hoshino F, Fukushi S, Shinohara M, Uchida K, Suzuki Y, Gojobori T, Takeda N. 2002. Phylogenetic analysis of the complete genome of 18 Norwalk-like viruses. *Virology* 299: 225–239. doi:10.1006/viro.2002.1568.

Keefe AD, Pai S, Ellington A. 2010. Aptamers as therapeutics. *Nat Rev Drug Discov* 9: 537–550. doi:10.1038/nrd3249.

Kitajima M, Wang N, Tay M, Miao J, Whittle A. 2016. Development of a MEMS-based electrochemical aptasensor for norovirus detection. *Micro Nano Lett* 11: 582–585. doi:10.1049/mnl.2016.0295.

Köhler G, Milstein C. 1975. Continuous cultures of fused cells secreting antibody of predefined specificity. *Nature* 256: 495–497. doi:10.1038/256495a0.

Kou B, Crawford SE, Ajami NJ, Czakó R, Neill FH, Tanaka TN, Kitamoto N, Palzkill TG, Estes MK, Atmar RL. 2015. Characterization of cross-reactive norovirus-specific monoclonal antibodies. *Clin Vaccine Immunol* 22: 160–167. doi:10.1128/CVI.00519-14.

Lee JH, Canny MD, De Erkenez A, Krilleke D, Ng YS, Shima DT, Pardi A, Jucker F. 2005. A therapeutic aptamer inhibits angiogenesis by specifically targeting the heparin binding domain of VEGF165. *Proc Natl Acad Sci USA* 102: 18902–18907. doi:10.1073/pnas.0509069102.

Mairal T, Cengiz Özalp V, Lozano Sánchez P, Mir M, Katakis I, O'Sullivan CK. 2008. Aptamers: Molecular tools for analytical applications. *Anal Bioanal Chem* 390: 989–1007. doi:10.1007/s00216-007-1346-4.

Miura T, Sano D, Suenaga A, Yoshimura T, Fuzawa M, Nakagomi T, Nakagomi O, Okabe S. 2013. Histo-blood group antigen-like substances of human enteric bacteria as specific adsorbents for human noroviruses. *J Virol* 87: 9441–9451. doi:10.1128/JVI.01060-13.

Moore MD, Bobay BG, Mertens B, Jaykus LA. 2016. Human norovirus aptamer exhibits high degree of target conformation-dependent binding similar to receptors and discriminates particle functionality. *mSphere* 1: pii-e00298-16 doi:10.1128/mSphere.00298-16.

Moore MD, Escudero-Abarca BI, Suh SH, Jaykus LA. 2015. Generation and characterization of nucleic acid aptamers targeting the capsid P domain of a human norovirus GII.4 strain. *J Biotechnol* 209: 41–49. doi:10.1016/j.jbiotec.2015.06.389.

Patel MM, Hall AJ, Vinjé J, Parashar UD. 2009. Noroviruses: A comprehensive review. *J Clin Virol* 44: 1-8. doi:10.1016/j.jcv.2008.10.009.

Prasad BV, Shanker S, Muhaxhiri Z, Deng L, Choi JM, Estes MK, Song Y, Palzkill T, Atmar RL. 2016. Antiviral targets of human noroviruses. *Curr Opin Virol* 18: 117–125. doi:10.1016/j.coviro.2016.06.002.

Prasad BV, Hardy ME, Dokland T, Bella J, Rossmann MG, Estes MK. 1999. X-ray crystallographic structure of the Norwalk virus capsid. *Science*. 286: 287-290.doi:10.1126/science.286.5438.287.

Robilotti E. 2015. Norovirus. *Clin Microbiol Rev* 28: 134–163.

Shum KT, Zhou J, Rossi JJ. 2013. Aptamer-based therapeutics: New approaches to combat human viral diseases. *Pharmaceuticals* 6: 1507–1542. doi:10.3390/ph6121507.

Song S, Wang L, Li J, Fan C, Zhao J. 2008. Aptamer-based biosensors. *Trends Anal Chem* 27: 108–117. doi:10.1016/j.trac.2007.12.004.

Stoltenburg R, Reinemann C, Strehlitz B. 2007. SELEX-A revolutionary method to generate high-affinity nucleic acid ligands. *Biomol Eng* 24: 381–403. doi:10.1016/j.bioeng.2007.06.001.

Tuerk C, Gold L. 1990. Systematic evolution of ligands by exponential enrichment: RNA ligands to bacteriophage T4 DNA polymerase. *Science* 249: 505–510. doi:10.1126/science.2200121.

Van den Kieboom CH, van der Beek SL, Meszaros T, Gyurcsanyi RE, Ferwerda G, de Jonge MI. 2015. Aptasensors for viral diagnostics. *Trends Anal Chem* 74: 58–67 doi:10.1016/j.trac.2015.05.012.

Vinjé J. 2015. Advances in laboratory methods for detection and typing of norovirus. *J Clin Microbiol* 53: 373–381. doi:10.1128/JCM.01535-14.

Winkler W, Nahvi A, Breaker RR. 2002. Thiamine derivatives bind messenger RNAs directly to regulate bacterial gene expression. *Nature* 419: 952–956. doi:10.1038/nature01145.

Wu Y, Zhong Z, Huber J, Bassi R, Finnerty B, Corcoran E, Li H, Navarro E, Balderes P, Jimenez X, Koo H, Mangalampalli VRM, Ludwig DL, Tonra JR, Hicklin DJ. 2006. Anti-vascular endothelial growth factor receptor-1 antagonist antibody as a therapeutic agent for cancer. *Clin Cancer Res* 12: 6573–6584. doi:10.1158/1078-0432.CCR-06-0831.

Yalow RS, Berson SA. 1959. Assay of plasma insulin in human subjects by immunological methods. *Nature* 184: 1648–1649. doi:10.1038/1841648b0.

In: Noroviruses
Editor: Jesús L. Romalde

ISBN: 978-1-53611-939-8

Chapter 9

THE JOURNEY TOWARDS AN EFFECTIVE NOROVIRUS VACCINE

Jesús L. Romalde*
Departamento de Microbiología y Parasitología,
CIBUS-Facultad de Biología, Universidade de Santiago de Compostela,
Santiago de Compostela, Spain

ABSTRACT

Noroviruses (NoV) are the leading cause of acute non-bacterial gastroenteritis worldwide. NoV are highly infectious, with as low as <10 to 100 viral particles causing infection. This group of non-enveloped, single-stranded RNA viruses, belonging to the *Caliciviridae* family, is highly resistant to environmental conditions. Of the different NoV genogroups described, genogroups I, II and, in a lesser extent IV, are the most frequently associated with both disease outbreaks and sporadic cases. The main routes of transmission include the fecal-oral route, aerosolized vomitus, person-to-person, and contaminated food, water and fomites. The lack of a realiable cell culture system or an animal model

* Corresponding Author address: Departamento de Microbiología y Parasitología, CIBUS-Facultad de Biología. Universidade de Santiago de Compostela. Campus Vida s/n. 15782, Santiago de Compostela. Spain. Email: jesus.romalde@usc.es.

has hampered the development of vaccines, and ongoing formulations are based on recombinant capsid proteins, such as virus-like particles (VLP) and P particles. These candidate NoV vaccines are showing promising results in terms of tolerability and immune response. This chapter will review the recent advances in NoV vaccine development.

Keywords: norovirus, vaccine, VLP, P particles, prevention

INTRODUCTION

Noroviruses (NoV) are the leading cause of non-bacterial acute gastroenteritis (Glass et al., 2009; Siebenga et al., 2009), causing approximately 50% of all such outbreaks. Although for most people the disease is relatively mild and self-limiting, morbidity and mortality are of concerns for children, immunocompromised individuals, and elderlies (Bok and Green, 2012; Harris et al., 2008; Okada et al., 2006; Payne et al., 2013; Siebenga et al., 2008; Trivedi et al., 2012). Symptoms comprise acute onset vomiting and watery diarrhea that commonly persist from one to three days (Rockx et al., 2002). It has been estimated that the prevalence of NoV among all cases of acute gastroenteritis is 18% (Ahmed et al., 2014) and that these viruses cause around 200,000 deaths per year worldwide (Lopman, 2015; Patel et al., 2009). In England, children under five years present the highest incidence, accounting for five times that of the general population (Phillips et al., 2009), and in USA the highest rates of emergency department, outpatients and inpatients visits among pediatric population occur due to NoV infections (Lopman et al., 2011; Gastañaduy et al., 2013). In fact, NoV replaced rotavirus as the most common cause of gastroenteritis in young children, thanks to the success of rotavirus vaccination in this age group (Koo et al., 2013; Payne et al., 2013). In the elderly population, mortality rates due to NoV placed in second place after those caused by *Clostridium difficile* (Hall et al., 2012; Trivedi et al., 2012; van Asten et al., 2012).

NoV are extremely contagious, owing to a very low infectious dose and to high levels of shedding (Chan et al., 2006; Teunis et al., 2008; Tu et al., 2008). Additionally, although patients usually recover from symptoms in 2-3 days, high levels of shedding may last several weeks (Tu et al., 2008). No or limited long-term immunity results from infection, so one person may be repeatedly infected (Johnson et al., 1990).

NOROVIRUS: CHARACTERISTICS AND PATHOGENESIS

Physical and Genetic Properties

Norovirus, classified in the family *Caliciviridae*, is a group of non-enveloped, icosahedral viruses with ~38 nm in diameter (Prassad et al., 1999). The NoV genome is a linear, non-segmented, single-stranded, positive sense RNA of approximately 7.5 kb in length containing three open reading frames (ORF1-ORF3)(Jiang et al., 1993; Lambden et al., 1993). The genome is protein-linked at the 5' end and polyadenylated at the 3' end.

ORF1 encodes a non-structural polyprotein precursor, which is cleaved into NS1/2 (p48), NS3 (nucleotide triphosphatase), NS4 (p22), NS5 (VPg), NS6 (protease) and NS7 (RNA-dependent RNA-polymerase). ORF2 encodes the major capsid protein (VP1) and ORF3 encodes the minor structural protein (VP2)(Hardy, 2005). VP1 and VP2 both are synthesized from a protein-linked subgenomic RNA containing both ORF2 and ORF3.

The norovirus virion is composed of 90 dimers of the major capsid protein VP1 and one or two copies of the minor structural protein VP2 (Hardy, 2005). VP1 consists of shell (S) and protruding (P) domains and determines the antigenicity of NoV (Hansman et al., 2006). A virion comprising a single major capsid protein is unique among animal viruses and more closely resembles the capsid composition of plant viruses. Function of protein VP2 is related to estabilization of capsid estructure (Bertolotti-Ciarlet et al., 2003).

Viral Diversity

NoV are genetically diverse, being classify in at least 33 genotypes within at least 6 genogroups (GI-GVI)(Green et al., 1994; Wang et al., 1994; Ando et al., 2000; Patel et al., 2009). Genogroups I, II and rarely IV are responsible for human infections (Zheng et al., 2010). Genogroups I and II are the most diverse and were further classified into nine (GI.1-GI.9) and 22 (GII.1-GII.22) genotypes (Vinjé, 2015).

Sequences of NoV strains show a great genetic divergence. Several studies have established that the sequences of the capsid gene can vary up to 60% among the different genogroups and up to 57% in human NoV (Green et al., 1995; Katayama et al., 2002; Vinjé et al., 2000). Such diversity is higher than those of other single-stranded RNA virus, indicating a faster evolution for NoV (Zheng et al., 2006).

Within the genogroup II, strains belonging to genotype II.4 are the most common cause of gastroenteritis outbreaks, including pandemics, and sporadic cases since the mid-1990s (Siebenga et al., 2009; Leon et al., 2011). However, in the last years the newly emerged GII.17 genotype has become the predominant strain in some parts of Asia (de Graaf et al., 2015, 2016).

The high incidence of NoV infections seems to be related to the emergence of new GII.4 variants that had evolved by genetic drift of the capsid gene or homologous recombination (Bull et al., 2011; Eden et al., 2013; Medici et al., 2014; Motomura et al., 2008; Sabriá et al., 2014; Siebenga et al., 2007). It has been reported that NoV GII.4 variant US95/96 was predominant during the winter of 1995–1996 (Vinjé et al., 1997), and that it was worldwide detected from outbreaks during the next 7 years. In 2002, the variant named Farmington Hills was detected and became predominant (Widdowson et al., 2004). The main feature of this strain was the insertion of 1 amino acid in the hypervariable region of the capsid (Widdowson et al., 2004). From 2002, a new GII.4 variant became predominant every 2 or 3 years. Thus, the 2002 strain was replaced in 2004 by the 2004 variant Bull et al., 2006), which was then replaced by the 2006a and 2006b variants. Both 2006 variants co-circulated during 2006 and 2007 (Tu et al., 2011). Further, a 2010 variant (New Orleans_2009 strain) was identified, replacing the 2006 variants and accounting for 60% of outbreaks (Vega et al., 2011). The 2012 variant (Sydney variant), that emerged in different regions worldwide, has been responsible for most outbreaks during the last years (Bennet et al., 2013; van Beek et al., 2013). The classification of these emerging GII.4 variants is based on >5% amino acid variation in the VP1 gene from other GII.4 strains (Bull et al., 2006; Zheng et al., 2010).

This pattern of continuous lineage turnover, referred to as epochal evolution, with emerging new variants replacing previously predominant circulating ones, is strongly evocative of what is observed in the molecular epidemiology of influenza A virus (IAV) (Lindesmith et al., 2008). Recently, a Japanese team developed a model to predict the changes in the genotype composition of circulating NoV strains (Suzuki et al., 2016). This fitness model, which takes into account the fitness effect of herd immunity, the genotype composition in the previous season, random genetic drift, environmental factors, is similar to that employed for influenza virus (Luksa and Lassig, 2014). Such a model, which has been already implemented in the norovirus forecasting system in Japan (Suzuki et al., 2016), can be an interesting tool for the periodical reformulation of norovirus vaccines accordingly to genotypes shifts in a certain area.

Transmission

Strains of genotype GII.4 have been responsible of approximately 80% of the reported outbreaks during the last decade (Hoa Tran et al., 2013), although the prevalence of the infecting genotypes varies between routes of transmission or human populations (Kroneman et al., 2008). Genotype GII.4 has been more often associated to person-to-person transmission, while other genotypes, including GI.3, GI.6, GI.7, GII.3 or GII.12 were preferentially associated to food-borne transmission (Verhoef et al., 2015). In general, and problably associated with the higher stability of GI over GII strains, GI transmission is more related to water-borne transmission

(Lysén et al., 2009). In fact, NoV outbreaks have been associated to contamination events of water in wells and recreational water settings (i.e., beaches, lakes and pools)(Hlavsa et al., 2015).

Different types of food, including shellfish and fresh products, constitute an important route for the spread of NoV (de Graaf et al., 2016). Foods can be contaminated with human NoV at any point from farm to fork. For instance, shellfish can be contaminated by fecal discharges in the coastal harvesting areas, and fresh vegetables and fruits by irrigation with contaminated sewage (DiCaprio et al., 2015; Manso et al., 2013; Verhoef et al., 2015). In addition, infected food handlers can contaminate those products during either harvesting or processing. The globalization of food production and trade has increased the potential risk of transmission of strains between different parts of the world (Polo et al., 2010; de Graaf et al., 2016). However, the possible global scale of food-borne NoV outbreaks is difficult to determine due to the partial epidemiological studies performed by individual countries. However, a retrospective study has shown that a percentage of such outbreaks could be part of international events with a common source (Verhoef et al., 2011).

Enclosed environments, such as cruise ships, detention centers, prisons, and health-care settings, favour the rapid transmission of NoV infections. In these environments, viral gastroenteritis, which occurs frequently and is highly contagious affecting residents and staff members, can be difficult to control due to the close living quarters, shared food and dining areas, etc. For instance, it has been reported that more than 90% of diarrheal outbreaks on cruise ships are caused by NoV (Duong et al., 2016). On the other hand, nosocomial transmission of NoV is also frequent, constituting an important burden for in-patient services (Ahmed et al., 2014). It has been reported that symptomatic shedders and/or immunocompromised patients chronically infected with NoV may act as reservoirs of the virus (Milbrath et al., 2013; Sukhrie et al., 2012).

Although no human infection with animal NoV strains has been reported until now, indirect evidences for transmission of NoV between different host species have been obtained. Thus, human NoV have been detected in fecal samples from pigs, dogs and cattle (Mattison et al., 2007;

Summa and von Bonsdorff, 2012) and, on the other hand, serum antibodies against bovine and canine noroviruses have been detected in humans (Widdowson et al., 2005; Mesquita et al., 2013).

Pathogenesis

Heterogeneous host-virus interactions characterize the pathogenesis of human NoV. Differences in susceptibility among individuals have been described, as well as differences in virulence among NoV genotypes (Friesema et al., 2009; Lindesmith et al., 2003; Thorven et al., 2005). Among other factors, the attachment to Histo-Blood Group Antigens (HBGA), the tissular tropism of the strains, the host immune responses, or the microbiota of the hosts have been described as playing important roles in NoV pathogenesis (de Graaf et al., 2016).

HBGA are highly polymorphic complex carbohydrates present in different cell types such as erythrocytes, vascular endothelial cells, gastrointestinal, urogenital and respiratory epithelial cells. In addition, it may be in saliva, milk and other body fluids depending on the secretory/non-secretory nature of the individual. HBGA are determinants of both the ABO blood group and Lewis Blood group systems (Stanley and Cummings, 2009). In certain cell types, α(1,2)-fucosyltransferase 2 (FUT2) adds a fucose group to precursors of HBGA generating H HBGA and subsequentely A and B HBGA. Individuals who lack FUT2 are known as non-secretors and A, B, and H HBGA are not present in their body fluids (Le Pendu et al., 2006). Although other receptors or coreceptors may also exist, NoV recognize and bind to HBGA as a pre-infection step with the exception of a few genotypes (de Graaf et al., 2016; Pringle et al., 2015). This recognition occurs in a very specific way. The major capsid protein VP1, contains sites that are predicted to interact with HBGA, namely the P2 variable region (Tan and Jiang, 2010), and therefore, the different strains of NoV have different profiles of binding to these receptors. In addition, mutations occurring in this region may be driven by evasion of the host immune response (Donaldson et al., 2008).

Because different strains of human NoV have different specificities for HBGA, most of them only infect a subset of the population. Even though several NoV strains show binding to a common HBGA, the affinity with which they bind may also vary (Huang et al., 2005; Lindesmith et al., 2008, 2010). All these facts suggest a pathogen-host co-evolution driven by viral protein-carbohydrate interactions (Le Pendu et al., 2006).

The possible implication of other factors, in addition to HBGA, in attachment and entry cannot be ruled out, and further studies are needed in order to clarify, if such factors exist, their role in norovirus pathogenesis (de Graaf et al., 2016).

Histochemical approaches have been employed to study the host cell tropism of NoV (Chan et al., 2011; Murakami et al., 2013; Yazawa et al., 2014), but discrepant results have been obtained from different authors. Thus, it has been suggested that NoV GII.6 and GII.I bind to epithelial and goblet cells in the ileum, or that GI.1, GII.2 and GII.6 VLP attach to the luminal epithelium of jejunal mucosae. Other studies revealed the binding of NoV GII.4 to cells of Brunner's glands and lamina propria, but not to cells of the luminal surface of the epithelium of duodenal tissue. If these different results reflect actual differences in host cell tropism among diverse NoV genotypes/strains require further investigation, as well as the possible influences on the NoV pathogenesis.

Some recent studies showed that bacteria in the host microbiota might influence human NoV infections (Jones et al., 2014). The mechanisms to this influence include the provision of additional HBGA for NoV binding, the modification of host HBGA expression, and the modulation of host immunity. It is known that several bacterial species usually found in the gut microbiota, including *Escherichia coli*, *Helicobacter pylori* or *Enterobacter cloacae*, are able to express HBGA-like sugars (Jones et al., 2014; Rasko et al., 2000; Yi et al., 2005). The adhesion of NoV to bacterial sugars may enhance the viral uptake into the host cells (Jones et al., 2014).

It is noteworthy to mention here that for years all attempts to propagate human NoV strains have been unsuccessful. More than 26 different cell lines had been evaluated combined with a variety of culture media and growth conditions but without cytopathic effect or replication (Duizer et al., 2004), reflecting the lack of awareness on the cell types that NoV infect *in vivo*. In 2007, Straub et al. showed for the first time that noroviruses infect and replicate in a physiologically relevant 3-dimensional (3-D), organoid model of human small intestinal epithelium. A successful *in vitro* passage of both GI and GII was achieved, although this work could not be confirmed in other laboratories. Very recently, Ettayebi et al. (2016) developed a culture system that mimic the human intestinal epithelium and supports the growth of NoV GII.4 and GII.3 strains. The approach consists in the generation of multicellular differentiated human intestinal enteroids from stem cells isolated from intestinal crypts in human intestinal tissues (Sato et al., 2011; Saxena et al., 2016). The human enteroids will allow deeper studies on host-pathogen interactions, as well as the assessment of control and treatment methods for NoV.

Vaccine Development

Noroviruses have been defined as the perfect human pathogens (Hall, 2012). As mentioned already in this chapter and throughout the entire book, they possess all the attributes of an ideal infectious agent, including high contagiousness, constant evolution, moderate virulence, evocation of limited immunity, and maintenance of a large susceptible pool of hosts. Since NoV infections are difficult to avoid and there is no etiologic treatment (Baehner et al., 2016), this agent represents a good candidate for potential prevention through vaccination.

The study of the NoV immunity and the development of vaccines have been hampered by the lack of efficient systems to propagate the virus. Therefore, such studies have been performed with nonreplicating recombinant proteins either as virus-like particles (VLP) or as P particles. The use of VLP in HBGA binding experiments as a surrogate of neutralization assays indicated that increased serum titres of antibodies that inhibited VLP binding to HBGA correlated with reductions in infection rates and severity of the disease (Lindsmith et al., 2015; Melhem, 2016). In addition, the presence of specific immunoglobulin A (IgA) in saliva and/or faeces and IgG was associated with protection against NoV infection (Lindesmith et al., 2003). All these findings support a role of the host immune responses in reducing the viral load, the duration of the virus shedding and the severity of the disease. The duration of immunity following infection could be also investigated and has been up to 6 months but less than 2 years in the last studies (Atmar and Estes, 2012).

Although there are currently no licensed vaccines for NoV, signifficant effort has led to the development of an efficient vaccine covering the major genotypes, namely GI.1 and GII.4 (Prasad et al., 2016; Riddle and Walker, 2016).

VLP-Based Vaccines

NoV VLP vaccines derived from various production approaches have been extensively evaluated in preclinical animal trials (Tan and Jiang, 2014), although currently baculovirus-expressed VLP are the most utilized. First studies in a mice model showed that intranasal administration of VLP stimulate NoV-specific IgG and mucosal IgA responses (Richardson et al., 2013; Ball et al., 1998). In addition, the use of immunostimulants induced higher IgA-secreting cells in Peyers patches and Th2 cytokines (Richardson et el., 2013). These studies provided the proof of concept that mucosal administration of VLP-based vaccines can generate protective response in animal models (Melhem, 2016). Other studies utilized different animal models, including gnotobiotic pigs or calves and chimpazees, to demonstrate that vaccinated animals with recombinant VLP were protected against homologous strains (Bok et al., 2011; Cohen, 2011; Souza et al., 2007, 2008). Velasques et al. (2011) tested a bivalent VLP GI.1/GII.4 formulation for intranasal use in various laboratory animals demonstrating robust systemic and mucosal immunity. More recently, Tamminen et al. (2016) observed that although intranasal and intramuscular immunization of Balb/c mice with NoV GII.4 VLP induce similar blocking activity in serum, only intranasal vaccination induces the development of functional anti-NoV IgA on mucosal surface, which strongly correlates with blocking activity in mucosa.

Following demonstrations of safety and immunogenicity in animals, the baculovirus-expressed NoV VLP were further studied in phase I and phase II trials for safety, immunogenicity and efficacy in healthy volunteers. Several human trials testing NoV vaccines have been performed during the last fifteen years. Many of these studies used VLP-based vaccines that proved safe and immunogenic when orally or intranasally administered (Ball et al., 1999; El-Kamary et al., 2010; Tacket et al., 2003). Atmar et al. (2011) conducted a randomized, double-blind, placebo-controlled trial at four clinical sites to evaluate the safety, immunogenicity and efficacy of an intranasally delivered NoV GI.1 VLP vaccine. NoV VLP were produced in a baculovirus expression system, using monophosphoryl lipid A as an adjuvant, chitosan as a mucoadhesive agent, and sucrose and mannitol excipients as bulking agents to stabilize the VLPs during lyophilization. A virus–specific IgA seroresponse was detected in 70% of vaccine recipients. In addition, vaccination significantly reduced the frequencies of NoV gastroenteritis and infection. Ramirez et al. (2012) administered different doses of the same formulation and observed significant dose-dependent IgA or IgG responses and a good correlation between the B memory (BM) cell frequencies and serum antibody levels and mucosally-primed antibody-secreting cell responses. These works demonstrated the ability of this VLP-vaccine to generate a strong humoral immune response with possible long-term protection.

All the previous mentioned studies were performed using a GI.1 strain, but GII.4 is the predominant NoV genotype throughout the world. On the other hand, human challenge studies have shown poor cross-reactivity between viruses from GI and GII (Malm et al., 2015; Pringle et al., 2015). The first study involving humans performed using a NoV GII.4 strain was carried out by Frenck et al. (2012) to evaluate the relationship between HBGA and NoV infection and associated illness. Secretor and non-secretor individuals were challenged with the NoV GII.4 strain Cin-1, originally isolated from human stool. Seventy percent of secretors were infected with

NoV and 57% became ill with vomiting and/or diarrhea, whereas only 1 non-secretor became ill, and another shed virus for a single day.

Due to the high number of circulating genetically diverse viruses and the annual variability of NoV genotypes, it seems crucial to develop a vaccine that offers heterologous protection. Thus, further studies focused on bivalent NoV formulations, including GI and GII recombinant VLP or GII.4 consensus VLP (based on the capsid sequences of 3 different pandemic variants). Parra et al. (2012) demonstrated that the consensus GII.4 vaccine induced the production of antibodies in intranasal or intramuscular vaccinated rabbits. Following these studies, these two candidate VLP vaccines have been moved to clinical trials for safety and efficacy through human volunteer challenge studies. Data from studies in adult volunteers showed that bivalent and multivalent VLP candidate vaccines induced a broad antibody response to multiple epitopes of vaccine and non-vaccine NoV strains (Lindesmith et al., 2015). Similar results were obtained in mice by Malm et al. (2015) with a bivalent vaccine containing GI.3 and GII.4 (1999 strain), although these authors advise that a carefully selection of the genotypes to be included in the formulation is needed.

The most recent studies included a intramuscular bivalent VLP vaccine (GI.1 plus GII.4 consensus), developed by Takeda Pharmaceutical Company Limited, which elicited robust and high immune responses in human trials, measured as total Ig, IgG and IgA, as well as functional antibody responses assessed by the ability to block binding to HBGA (Bernstein et al., 2015; Lindesmith et al., 2015; Treanor et al., 2014). This vaccine, which is in phase 2 clinical trials, contains monophosphoryl lipid A and aluminum as adjuvants (Baehner et al., 2016). On the other hand, Atmar et al. (2016) comparatively evaluated in healthy adults two different formulations of this vaccine, containing different proportions of GI.1 (15 or 50μg per dose) and GII.4 antigens (50μg per dose). Both formulations rendered rapid immune responses to vaccination although genotype specific response varied depending on the antigen dose received. Further studies are in progress on the usefulness of a booster dose or the necessity of the monophosphoryl lipid A as adjuvant (Atmar et al., 2016).

P Particles-Based Formulations

As mentioned before, capsid protein VP1 contains two major domains, shell (S) and protruding (P) domains, linked by a flexible hinge (Prasad et al., 1999). Expression f P domain in *E. coli* yields variable complexes, including P dimers, smaller P particles (containing 6 dimers), P particles (containing 12 dimers) and polyvalent P complexes (containing many P dimers), which retain the ability to bind to HBGA ligands (Tan and Jiang, 2014).

P particles, made by expression of the VP1 P domain with end-terminal cysteine residues in a prokariotic vector, are highly stable and highly immunogenic and have been appealed as vaccine alternatives to VLP (Fang et al., 2103; Tan and Jiang, 2014; Tan et al., 2008, 2011). P particles contain the P1 domain in the inner core and the P2 binding domain in the outer layer, and present the same binding profile as the native capsid and VLP (Tan and Jiang, 2005).

Studies on immunogenicity and protective efficacy of P particles are more limited than those mentioned for VLP. Tamminen et al. (2012) suggested that P particles were less immunogenic than VLP, but these results are in contradiction with those from other authors (Tan and Jiang, 2012) who claim concerns that the former study was performed with P dimers instead of P particles. In fact, Fang et al. (2013) demonstarted that P particles are more immunogenic than P dimers and stimulate immune response in mice similar to VLP. Recent studies performed by Verma and coworkers (2016) showed enhanced immune responses induced by recombinant NoV P dimer vaccines using the *Vibrio vulnificus* flagellin protein FlaB as adjuvant. Different studies in animal models using diverse vaccine regimes and/or doses indicated that P particles confer similar or even higher protective efficacy than VLP formulations with the same regime (Kocher et al., 2014).

The polyvalent P domain complexes, made by fusion of two or three NoV P domains, provide opportunities to develop and formulate dual or multivalent vaccines against different GI and GII genotypes (Wang et al., 2013), that could offer broad protection against NoV infections.

Polyvalent Vaccines

Blazevic et al. (2011) developed a bivalent vaccine against NoV and Rotavirus (RV), including NoV VLP and RT VP6, and used it to immunize Balb/c mice. The results showed that the two antigens were highly immunogenic in a mixture, but also that RV VP6 may have the potential to act as an adjuvant for the NoV VLP vaccine. Later, Malm et al. (2016) proved also that RV VP6 upregulated the expression of cell surface activation and maturation molecules and induced secretion of several proinflamatory cytokines and chemokynes. In addition, the coadministration of VP6 with GII.4 VLP improved the uptake of the later to RAW 264.7 cells, which further supported the use of VP6 as a part of nonlive RV-NoV combination vaccine.

P particles can also be used as a platform for expression of oher viral antigens (Kocher and Yuan, 2015), and this capacity was leveraged to develop polyvalent vaccines (Tan and Jiang, 2014). The first P particle chimeric vaccine contained a surface insertion of the RV neutralizing antigen VP8* (Tan et al., 2011) and induced strong immune responses and protection aginst both viruses in mice. Another dual vaccine was developed containing the P particle the conserved M2e epitope of influenza A viruses (Xia et al., 2011). This formulation yielded high titers of specific antibodies against both RV M2e epitope and NoV P domain and fully protected mice in challenges with lethal doses of influenza viruses.

On the other hand, Wang et al., (2014) developed a dual vaccine candidate against NoV and hepatitis E virus (HEV) based on large polyvalent protein complexes including the fused dimeric P domain of both viruses. Recombinant proteins, expressed in *E. coli* as a construct glutathione-s-transferase (GST) tagged, were employed to immunize mice. Induction of high antibody titers, high neutralizing activity and ability to block the HBGA-binding was observed. Recently, Xia et al. (2016) proposed a trivalent subunit vaccine, consisting in a recombination fusion protein containing the P domains of an Asian Astrovirus (AstV), a zoonotic G3 HEV and a human NoV GII.4. This trivalent vaccine elicited high antibody responses against the three viruses in mice, as well as HEV neutralizing titers and blocking activity against NoV binding to HBGA ligands, which support the authors' claim as a promising vaccine candidate against HEV, AstV and NoV.

Conclusion

NoV pose a significant global health concern. The intense effort for vaccine development in the last years achieved encouraging results as immunogenic VLP and P particle candidate vaccines, some of them in phase II clinical trials, or the designing of multivalent vaccines to counter the antigenic and genetic diversity of NoV. Further studies with those candidate vaccines are needed on the determination of vaccine regimes, doses, administration routes or duration of immunity. Furthermore, increased knowledge on the worldwide circulating strains, the existence of shared antigenic groups among NoV genotypes and the development of models for predicting genotype composition for NoV seasons are essential to develop a globally effective norovirus vaccine.

Acknowledgment

The critical reading of the manuscript by Dr. S. Balboa is gratefully acknowledged.

References

Ahmed SM, Hall AJ, Robinson AE, Verhoef L, Premkumar P, Parashar UD, Koopmans M, Lopman BA. 2014. Global prevalence of norovirus in cases of gastroenteritis: a systematic review and meta-analysis. *Lancet Infect Dis.* 14: 725-730.

Ando T, Noel JS, Fankhauser RL. 2000. Genetic classification of "Norwalk-like viruses." *J Infect Dis* 181: 336-348.

Atmar RL, Estes MK. 2012. Norovirus vaccine development: next steps. *Expert Rev. Vaccines* 11: 1023-1025

Atmar RL, Bernstein DI, Harro CD, Al-Ibrahim MS, Chen WH, Ferreira J, Estes MK, Graham DY, Opekun AR, Richardson C, Mendelman PM. 2011. Norovirus vaccine against experimental human Norwalk Virus illness. *N Engl J Med* 365: 2178-2187.

Atmar RL, Baehner F, Cramer JP, Song E, Borkowski A, Mendelman PM, NOR-201 Study Group. 2016. Rapid Responses to 2 Virus-Like Particle Norovirus Vaccine Candidate Formulations in Healthy Adults: A Randomized Controlled Trial. *J Infect Dis* 214: 845-853.

Baehner F, Bogaerts H, Goodwin R. 2016. Vaccines against norovirus: state of the art trials in children and adults. *Clin Microbiol Infect* 22: S136-S139.

Ball JM, Hardy ME, Atmar RL, Conner ME, Estes MK. 1998. Oral immunization with recombinant Norwalk virus-like particles induces a systemic and mucosal immune response in mice. *J Virol* 72: 1345-1353.

Ball JM, Graham DY, Opekun AR, Gilger MA, Guerrero RA, Estes MK. 1999. Recombinant Norwalk virus-like particles given orally to volunteers: phase I study. *Gastroenterology* 117: 40-48.

Bennett S, MacLean A, Miller RS, Aitken C, Gunson RN. 2013. Increased norovirus activity in Scotland in 2012 is associated with the emergence of a new norovirus GII.4 variant. Euro Surveill 18: pii=20349.

Bernstein DI, Atmar RL, Lyon GM, Treanor JJ, Chen WH, Jiang X, Vinjé J, Gregoricus N, Frenck RW Jr, Moe CL, Al-Ibrahim MS, Barrett J, Ferreira J, Estes MK, Graham DY, Goodwin R, Borkowski A, Clemens R, Mendelman PM. 2015. Norovirus vaccine against experimental human GII.4 virus illness: a challenge study in healthy adults. *J Infect Dis* 211: 870-878.

Bertolotti-Ciarlet A, White LJ, Chen R, Prasad BV, Estes MK. 2003. Structural requirements for the assembly of Norwalk virus-like particles. *J Virol* 76: 4044-4055.

Blazevic V, Lappalainen S, Nurminen K, Huhti L, Vesikari T. 2011. Norovirus VLPs and rotavirus VP6 protein as combined vaccine for childhood gastroenteritis. *Vaccine* 29: 8126-8133.

Bok K, Green KY. 2012. Norovirus gastroenteritis in immune compromised patients. *N Engl J Med.* 367: 2126-2132.

Bok K, Parra GI, Mitra T, Abente E, Shaver CK, Boon D, Engle R, Yu C, Kapikian AZ, Sosnovtsev SV, Purcell RH, Green KY. 2011. Chimpanzees as an animal model for human norovirus infection and vaccine development. *Proc Natl Acad Sci USA* 108: 325-330.

Bull RA, White PA. 2011. Mechanisms of GII.4 norovirus evolution. *Trends Microbiol* 19: 233-240.

Bull RA, Tu ET, McIver CJ, Rawlinson WD, White PA. 2006. Emergence of a new norovirus genotype II.4 variant associated with global outbreaks of gastroenteritis. *J Clin Microbiol* 44: 327-333.

Chan MC, Sung JJ, Lam RK, Chan PK, Lee NL, Lai RW, Leung WK. 2006. Fecal viral load and norovirus-associated gastroenteritis. *Emerg Infect Dis* 12: 1278-1280.

Chan MC, Ho WS, Sung JJ. 2011. In vitro whole-virus binding of a norovirus genogroup II genotype 4 strain to cells of the lamina propria and Brunner's glands in the human duodenum. *J Virol* 85: 8427-8430.

Cohen J. 2011. Committee recommends NIH continue chimpanzee research, with new limits. *Science* 334: 1484-1485

de Graaf M, van Beek J, Vennema H, Podkolzin AT, Hewitt J, Bucardo F, Templeton K, Mans J, Nordgren J, Reuter G, Lynch M, Rasmussen LD, Iritani N, Chan MC, Martella V, Ambert-Balay K, Vinjé J, White PA, Koopmans MP. 2015. Emergence of a novel GII.17 norovirus – End of the GII.4 era? *Euro Surveill* 20: pii: 21178.

de Graaf M, van Beek J, Koopmans MPG. 2016. Human norovirus transmission and evolution in a changing world. *Nat Rev Microbiol* 14: 421-433.

DiCaprio E, Culbertson D, Li J. 2015. Evidence of the internalization of animal caliciviruses via the roots of growing strawberry plants and dissemination to the fruit. *Appl Environ Microbiol* 81: 2727-2734.

Donaldson EF, Lindesmith LC, Lobue AD, Baric RS. 2008. Norovirus pathogenesis: mechanisms of persistence and immune evasion in human populations. *Immunol Rev* 225: 190-211.

Duizer E, Schwab KJ, Neill FH, Atmar RL, Koopmans MP, Estes MK. 2004. Laboratory efforts to cultivate noroviruses. *J Gen Virol* 85: 79-87.

Duong KK, Mitruka K, Lippold SA. 2016. Cruise ship travel. In: Brunette GW (ed.) *CDC Health Information for International Travel.* Oxford University Press, NY.

Eden JS, Tanaka MM, Boni MF, Rawlinson WD, White PA. 2013. Recombination within the pandemic norovirus GII.4 lineage. *J Virol* 87: 6270-6282.

El-Kamary SS, Pasetti MF, Mendelman PM, Frey SE, Bernstein DI, Treanor JJ, Ferreira J, Chen WH, Sublett R, Richardson C, Bargatze RF, Sztein MB, Tacket CO. 2010. Adjuvanted intranasal Norwalk virus-like particle vaccine elicits antibodies and antibody-secreting cells that express homing receptors for mucosal and peripheral lymphoid tissues. *J Infect Dis* 202: 1649-1658.

Ettayebi K, Crawford SE, Murakami K, Broughman JR, Karandikar U, Tenge VR, Neill FH, Blutt SE, Zeng XL, Qu L, Kou B, Opekun AR, Burrin D, Graham DY, Ramani S, Atmar RL, Estes MK. 2016. Replication of human noroviruses in stem cell-derived human enteroids. *Science* 353: 1387-1393.

Fang H, Tan M, Xia M, Wang L, Jiang X. 2013. Norovirus P particle efficiently elicits innate, humoral and cellular immunity. *PLoS One* 8: e63269.

Frenck R, Bernstein DI, Xia M, Huang P, Zhong W, Parker S, Dickey M, McNeal M, Jiang X. 2012. Predicting susceptibility to norovirus GII.4 by use of a challenge model involving humans. *J Infect Dis* 206: 1386-1393.

Friesema IH, Vennema H, Heijne JC, de Jager CM, Teunis PF, van der Linde R, Duizer E, van Duynhoven YT. 2009. Differences in clinical presentation between norovirus genotypes in nursing homes. *J Clin Virol* 46: 341-344.

Gastañaduy PA, Vicuña Y, Salazar F, Broncano N, Gregoricus N, Vinjé J, Chico M, Parashar UD, Cooper PJ, Lopman B. 2015. Transmission of Norovirus within Households in Quininde, Ecuador. *Pediatr Infect Dis J.* 34: 1031-3.

Glass RI, Parashar UD, Estes MK. 2009. Norovirus gastroenteritis. *N Engl J Med* 361: 1776-1785.

Green SM, Dingle KE, Lambden PR, Caul EO, Ashley CR, Clarke IN. 1994. Human enteric *Caliciviridae*: A new prevalent small roundstructured virus group defined by RNA-dependent RNA polymerase and capsid diversity. *J Gen Virol* 75: 1883-1888.

Green SM, Lambden PR, Caul EO, Ashley CR, Clarke IN. 1995. Capsid diversity in small round-structured viruses: molecular characterization of an antigenically distinct human enteric calicivirus. *Virus Res* 37: 271-283.

Hall AJ. 2012. Noroviruses: The Perfect Human Pathogens? *J Infect Dis* 205: 1622-1624.

Hall AJ, Curns AT, McDonald LC, Parashar UD, Lopman BA. 2012. The roles of *Clostridium difficile* and norovirus among gastroenteritis-associated deaths in the United States, 1999-2007. *Clin Infect Dis* 55: 216-23.

Hlavsa MC, Roberts VA, Kahler AM, Hilborn ED, Mecher TR, Beach MJ, Wade TJ, Yoder JS. 2015. Outbreaks of Illness Associated with Recreational Water — United States, 2011–2012. MMWR 64: 668-672.

Hardy ME. 2005. Norovirus protein structure and function. *FEMS Microbiol Lett* 253: 1-8.

Harris JP, Edmunds WJ, Pebody R, Brown DW, Lopman BA. 2008. Deaths from norovirus among the elderly, England and Wales. *Emerg Infect Dis* 14: 1546-1552.

Hoa Tran TN, Trainor E, Nakagomi T, Cunliffe NA, Nakagomi O. 2013. Molecular epidemiology of noroviruses associated with acute sporadic gastroenteritis in children: global distribution of genogroups, genotypes and GII.4 variants. *J Clin Virol* 56: 185-193.

Huang P, Farkas T, Zhong W, Tan M, Thornton S, Morrow AL, Jiang X. 2005. Norovirus and histo-blood group antigens: demonstration of a wide spectrum of strain specificities and classification of two major binding groups among multiple binding patterns. *J Virol* 79: 6714-6722.

Jiang X, Wang M, Wang K, Estes MK. 1993. Sequence and genomic organization of Norwalk virus. *Virology*. 195: 51-61.

Johnson PC, Mathewson JJ, DuPont HL, Greenberg HB. 1990. Multiple-challenge study of host susceptibility to Norwalk gastroenteritis in US adults. *J Infect Dis* 161: 18-21.

Jones MK, Watanabe M, Zhu S, Graves CL, Keyes LR, Grau KR, Gonzalez-Hernandez MB, Iovine NM, Wobus CE, Vinjé J, Tibbetts SA, Wallet SM, Karst SM. 2014. Enteric bacteria promote human and mouse norovirus infection of B cells. *Science* 346: 755-759.

Katayama K, Shirato-Horikoshi H, Kojima S, Kageyama T, Oka T, Hoshino F, Fukushi S, Shinohara M, Uchida K, Suzuki Y, Gojobori T, Takeda N. 2002. Phylogenetic analysis of the complete genome of 18 Norwalk-like viruses. *Virology* 299: 225-239.

Kocher J, Yuan L. 2015. Norovirus vaccines and potential antinorovirus drugs: recent advances and future perspectives. *Future Virol* 10: 899-913.

Kocher J, Bui T, Giri-Rachman E, Wen K, Li G, Yang X, Liu F, Tan M, Xia M, Zhong W, Jiang X, Yuan L. 2014. Intranasal P particle vaccine provided partial cross-variant protection against human GII.4 norovirus diarrhea in gnotobiotic pigs. *J Virol* 88: 9728-9743.

Koo HL, Neill FH, Estes MK, Munoz FM, Cameron A, DuPont HL, Atmar RL. 2013. Noroviruses: The most common pediatric viral enteric pathogen at a large university hospital after introduction of rotavirus vaccination. *J Pediatric Infect Dis Soc.* 2: 57-60.

Kroneman A, Verhoef L, Harris J, Vennema H, Duizer E, van Duynhoven Y, Gray J, Iturriza M, Böttiger B, Falkenhorst G, Johnsen C, von Bonsdorff CH, Maunula L, Kuusi M, Pothier P, Gallay A, Schreier E, Höhne M, Koch J, Szücs G, Reuter G, Krisztalovics K, Lynch M, McKeown P, Foley B, Coughlan S, Ruggeri FM, Di Bartolo I, Vainio

K, Isakbaeva E, Poljsak-Prijatelj M, Grom AH, Mijovski JZ, Bosch A, Buesa J, Fauquier AS, Hernandéz-Pezzi G, Hedlund KO, Koopmans M. 2008. Analysis of integrated virological and epidemiological reports of norovirus outbreaks collected within the Foodborne Viruses in Europe network from 1 July 2001 to 30 June 2006. *J Clin Microbiol* 46: 2959-2965.

Lambden PR, Caul EO, Ashley CR, Clarke IN. 1993. Sequence and genome organization of a human small round-structured (Norwalk-like) virus. *Science* 259: 516-519.

Le Pendu J, Ruvoën-Clouet N, Kindberg E, Svensson L. 2006. Mendelian resistance to human norovirus infections. *Semin Immunol* 18: 375-386.

Leon JS, Kingsley DH, Montes JS, Richards GP, Lyon GM, Abdulhafid GM, Fernandez ML, Teunis PF, Flick GJ, Moe CL. 2011. Randomized, double-blinded clinical trial for human norovirus inactivation in oysters by high hydrostatic pressure processing. *Appl Environ Microbiol* 77: 5476-5482.

Lindesmith L, Moe C, Marionneau S, Ruvoen N, Jiang X, Lindblad L, Stewart P, LePendu J, Baric R. 2003. Human susceptibility and resistance to Norwalk virus infection. *Nat Med* 9: 548-553.

Lindesmith LC, Donaldson EF, Lobue AD, Cannon JL, Zheng DP, Vinje J, Baric RS. 2008. Mechanisms of GII.4 Norovirus Persistence in Human Populations. *PLoS Med* 5: e31.

Lindesmith LC, Donaldson E, Leon J, Moe CL, Frelinger JA, Johnston RE, Weber DJ, Baric RS. 2010. Heterotypic humoral and cellular immune responses following Norwalk virus infection. *J Virol* 84: 1800-1815.

Lindesmith LC, Ferris MT, Mullan CW, Ferreira J, Debbink K, Swanstrom J, Richardson C, Goodwin RR, Baehner F, Mendelman PM, Bargatze RF, Baric RS. 2015. Broad blockade antibody responses in human volunteers after immunization with a multivalent norovirus VLP candidate vaccine: immunological analyses from a phase I clinical trial. *PLoS Med* 12: e1001807.

Lopman BA. 2015. Global burden of norovirus and prospects for vaccine development. Bill & Melinda Gates Foundation and CDC Foundation Report. 46 pp.

Lopman BA, Hall AJ, Curns AT, Parashar UD. 2011. Increasing rates of gastroenteritishospital discharges in US adults and the contribution of norovirus, 1996-2007. *Clin Infect Dis* 52: 466-472.

Luksza M, Lassig M. 2014. A predictive fitness model for influenza. *Nature* 507: 57-61.

Lysén M, Thorhagen M, Brytting M, Hjertqvist M, Andersson Y, Hedlund KO. 2009. Genetic diversity among food-borne and waterborne norovirus strains causing outbreaks in Sweden. *J Clin Microbiol* 47: 2411-2418.

Malm M, Tamminen K, Lappalainen S, Uusi-Kerttula H, Vesikari T, Blazevic V. 2015. Genotype considerations for virus-like particle-based bivalent norovirus vaccine composition. *Clin Vaccine Immunol* 22: 656-663.

Malm M, Tamminen K, Lappalainen S, Vesikari T, Blazevic V. 2016. Rotavirus recombinant VP6 nanotubes act as an immunomodulator and delivery vehicle for norovirus virus-like particles. *J Immunol Res* 2016:9171632.

Manso CF, Romalde JL. 2013. Role of Norovirus in acute gastroenteritis development in the northwest of Spain during 2010-2011 period. *J Med Virol* 85: 2009–2015.

Mattison K, Shukla A, Cook A, Pollari F, Friendship R, Kelton D, Bidawid S, Farber JM. 2007. Human noroviruses in swine and cattle. *Emerg Infect Dis* 13: 1184-1188.

Medici MC, Tummolo F, Martella V, Giammanco GM, De Grazia S, Arcangeletti MC, De Conto F, Chezzi, C, Calderaro A. 2014. Nover recombinant GII.P16_GII.13 and GII.P16_GII.3 norovirus strains in Italy. *Virus Res* 188: 142-145.

Melhem NM. 2016. Norovirus vaccines: Correlates of protection, challenges and limitations. *Hum Vaccin Immunother* 12: 1653-1669.

Mesquita JR, Costantini VP, Cannon JL, Lin SC, Nascimento MS, Vinjé J. 2013. Presence of antibodies against genogroup VI norovirus in humans. *Virol J.* 10: 176.

Milbrath MO, Spicknall IH, Zelner JL, Moe CL, Eisenberg JN. 2013. Heterogeneity in norovirus shedding duration affects community risk. *Epidemiol Infect* 141: 1572-1584.

Motomura K, Oka T, Yokoyama M, Nakamura H, Mori H, Ode H, Hansman GS, Katayama K, Kanda T, Tanaka T, Takeda N, Sato H, Norovirus Surveillance Group of Japan. 2008. Identification of monomorphic and divergent haplotypes in the 2006-2007 norovirus GII/4 epidemic population by genomewide tracing of evolutionary history. *J Virol* 82: 11247-11262.

Murakami K, Kurihara C, Oka T, Shimoike T, Fujii Y, Takai-Todaka R, Park Y, Wakita T, Matsuda T, Hokari R, Miura S, Katayama K. 2013. Norovirus binding to intestinal epithelial cells is independent of histo-blood group antigens. *PLoS One* 8: e66534.

Okada M, Tanaka T, Oseto M, Takeda N, Shinozaki K. 2006. Genetic analysis of noroviruses associated with fatalities in healthcare facilities. *Arch Virol* 151: 1635-1641.

Parra GI, Bok K, Taylor R, Haynes JR, Sosnovtsev SV, Richardson C, Green KY. 2012. Immunogenicity and specificity of norovirus Consensus GII.4 virus-like particles in monovalent and bivalent vaccine formulations. *Vaccine* 30: 3580-3586.

Patel MM, Hall AJ, Vinjé J, Parashar UD. 2009. Noroviruses: a comprehensive review. *J Clin Virol* 44: 1-8.

Payne DC, Vinjé J, Szilagyi PG, Edwards KM, Staat MA, Weinberg GA, Hall CB, Chappell J, Bernstein DI, Curns AT, Wikswo M, Shirley SH, Hall AJ, Lopman B, Parashar UD. 2013. Norovirus and medically attended gastroenteritis in U.S. children. *N Engl J Med* 368:1121-1130.

Phillips G, Lopman, B, Tam CC, Iturriza-Gomara M, Brown, D, Gary J. 2009. Diagnosing norovirus-associated infectious intestinal disease using viral load. *BMC Infect Dis* 9: 63.

Polo D, Vilariño ML, Manso CF, Romalde JL. 2010. Imported mollusks and dissemination of human enteric viruses. *Emerg Infect Dis* 16: 1036-1038.

Prasad BV, Hardy ME, Dokland T, Bella J, Rossmann MG, Estes MK. 1999. X-ray crystallographic structure of the Norwalk virus capsid. *Science.* 286: 287-290.

Prasad BV, Shanker S, Muhaxhiri Z, Deng L, Choi JM, Estes MK, Song Y, Palzkill T, Atmar RL. 2016. Antiviral targets of human noroviruses. *Curr Opin Virol* 18: 117-125.

Pringle K, Lopman B, Vega E, Vinjé J, Parashar UD, Hall AJ. 2015. Noroviruses: epidemiology, immunity and prospects for prevention. *Future Microbiol* 10: 53-67.

Ramirez K, Wahid R, Richardson C, Bargatze RF, El-Kamary SS, Sztein MB, Pasetti MF. 2012. Intranasal vaccination with an adjuvanted Norwalk virus-like particle vaccine elicits antigen-specific B memory responses in human adult volunteers. *Clin Immunol* 144: 98-108.

Rasko DA, Wang G, Monteiro MA, Palcic MM, Taylor DE. 2000. Synthesis of mono- and di-fucosylated type I Lewis blood group antigens by *Helicobacter pylori. Eur J Biochem* 267: 6059-6066.

Richardson C, Bargatze RF, Goodwin R, Mendelman PM. 2013. Norovirus virus-like particle vaccines for the prevention of acute gastroenteritis. *Expert Rev Vaccines* 12: 155-167.

Riddle MS, Walker RI. 2016. Status of vaccine research and development for norovirus. *Vaccine* 34: 2895-2899.

Rockx B, De Wit M, Vennema H, Vinje J, De Bruin E, Van Duynhoven Y, Koopmans M. 2002. Natural history of human calicivirus infection: a prospective cohort study. *Clin Infect Dis* 35: 246-253.

Sabriá A, Pintó RM, Bosch A, Bartolomé R, Cornejo T, Torner N, Martínez A, de Simón M, Domínguez A, Guix S, Catalan Viral Gastroenteritis Study Group. 2014. Molecular and clinical epidemiology of norovirus outbreaks in Spain during the emergence of GII.4 2012 variant. *J Clin Virol* 60: 96-104.

Sato T, Stange DE, Ferrante M, Vries RG, van Es JH, van den Brink S, van Houdt WJ, Pronk A, van Gorp J, Siersema PD, Clevers H. 2011. Long-term expansion of epithelial organoids from human colon, adenoma, adenocarcinoma, and Barrett's epithelium. *Gastroenterology* 141: 1762–1772.

Saxena K, Simon LM, Zeng XL, Blutt SE, Crawford SE, Sastri NP, Karandikar UC, Ajami NJ, Zachos NC, Kovbasnjuk O, Donowitz M, Conner ME, Shaw CA, Estes MK. 2016. A paradox of transcriptional and functional innate interferon responses of human intestinal enteroids to enteric virus infection. Proc Natl Acad Sci USA pii: 201615422.

Siebenga JJ, Beersma MF, Vennema H, van Biezen P, Hartwig NJ, Koopmans M. 2008. High Prevalence of Prolonged Norovirus Shedding and Illness among Hospitalized Patients: A Model for In Vivo Molecular Evolution. *J Infect Dis,* 198: 994-1001.

Siebenga JJ, Vennema H, Renckens B, de Bruin E, van der Veer B, Siezen RJ, Koopmans M. 2007. Epochal evolution of GGII.4 norovirus capsid proteins from 1995 to 2006. *J Virol* 81: 9932-9941.

Siebenga JJ, Vennema H, Zheng DP, Vinjé J, Lee BE, Pang XL, Ho EC, Lim W, Choudekar A, Broor S, Halperin T, Rasool NB, Hewitt J, Greening GE, Jin M, Duan ZJ, Lucero Y, O'Ryan M, Hoehne M, Schreier E, Ratcliff RM, White PA, Iritani N, Reuter G, Koopmans M. 2009. Norovirus illness is a global problem: emergence and spread of norovirus GII.4 variants, 2001-2007. *J Infect Dis* 200: 802-812.

Souza M, Costantini V, Azevedo MS, Saif LJ. 2007. A human norovirus-like particle vaccine adjuvanted with ISCOM or mLT induces cytokine and antibody responses and protection to the homologous GII.4 human norovirus in a gnotobiotic pig disease model. Vaccine 25: 8448-8459.

Souza M, Azevedo MS, Jung K, Cheetham S, Saif LJ. 2008. Pathogenesis and immune responses in gnotobiotic calves after infection with the genogroup II.4-HS66 strain of human norovirus. *J Virol* 82: 1777-1786.

Stanley P, Cummings RD. 2009. Structures common to different glycans. In: Varki A, Cummings RD, Esko JD, Freeze HH, Stanley P, Bertozzi CR, Hart GW, Etzler ME (eds.) *Essentials of Glycobiology,* 2nd ed., pp- 175-198. Cold Spring Harbor, NY.

Sukhrie FH, Teunis P, Vennema H, Copra C, Thijs Beersma MF, Bogerman J, Koopmans M. 2012. Nosocomial transmission of norovirus is mainly caused by symptomatic cases. *Clin Infect Dis* 54: 931-937.

Suma M, von Bonsdorff CH, Maunula L. Pet dogs – a transmission route for human noroviruses?. *J Clin Virol* 53: 244-247.

Suzuki Y, Doan YH, Kimura H, Shinomiya H, Shirabe K, Katayama K. 2016. Predicting genotype compositions in norovirus seasons in Japan. Microbiol Immunol 60: 418-426.Teunis PF, Moe CL, Liu P, Miller SE, Lindesmith L, Baric RS, Le Pendu J, Calderon RL. 2008. Norwalk virus: how infectious is it? *J Med Virol* 80: 1468-1476.

Tacket CO, Sztein MB, Losonsky GA, Wasserman SS, Estes MK. 2003. Humoral, mucosal, and cellular immune responses to oral Norwalk virus-like particles in volunteers. *Clin Immunol* 108: 241-247.

Tamminen K, Huhti L, Koho T, Lappalainen S, Hytönen VP, Vesikari T, Blazevic V. 2012. A comparison of immunogenicity of norovirus GII-4 virus-like particles and P-particles. *Immunology* 135: 89-99.

Tamminen K, Malm M, Vesikari T, Blazevic V. 2016. Mucosal Antibodies Induced by Intranasal but Not Intramuscular Immunization Block Norovirus GII.4 Virus-Like Particle Receptor Binding. *Viral Immunol* 29: 315-319.

Tan M, Jiang X. 2005. The P domain of norovirus capsid protein forms a subviral particle that binds to histo-blood group antigen receptors. *J Virol* 79: 14017-14030.

Tan M, Jiang X. 2012. Norovirus P particle: a subviral nanoparticle for vaccine development against norovirus, rotavirus and influenza virus. *Nanomedicine (Lond)* 7: 889-897.

Tan M, Jiang X. 2010. Norovirus gastroenteritis, carbohydrate receptors, and animal models. *PLoS Pathog* 6: e1000983.

Tan M, Jiang X. 2014. Vaccine against norovirus. *Hum Vaccin Immunother.* 10: 1449-1456.

Tan M, Fang P, Chachiyo T, Xia M, Huang P, Fang Z, Jiang W, Jiang X. 2008. Noroviral P particle: structure, function and applications in virus-host interaction. *Virology* 382: 115-123.

Tan M, Huang P, Xia M, Fang PA, Zhong W, McNeal M, Wei C, Jiang W, Jiang X. 2011. Norovirus P particle, a novel platform for vaccine development and antibody production. *J Virol* 85: 753-764.

Thorven M1, Grahn A, Hedlund KO, Johansson H, Wahlfrid C, Larson G, Svensson L. 2005. A Homozygous Nonsense Mutation (428G→A) in the Human Secretor (*FUT2*) Gene Provides Resistance to Symptomatic Norovirus (GGII) Infections. *J Virol* 79: 15351-15355.

Treanor JJ, Atmar RL, Frey SE, Gormley R, Chen WH, Ferreira J, Goodwin R, Borkowski A, Clemens R, Mendelman PM. 2014. A novel intramuscular bivalent norovirus virus-like particle vaccine candidate--reactogenicity, safety, and immunogenicity in a phase 1 trial in healthy adults. *J Infect Dis* 210: 1763-1771.

Trivedi TK, DeSalvo T, Lee L, Palumbo A, Moll M, Curns A, Hall AJ, Patel M, Parashar UD, Lopman BA. 2012. Hospitalizations and mortality associated with norovirus outbreaks in nursing homes, 2009-2010. *JAMA* 308: 1668-1675.

Tu ET, Bull RA, Greening GE, Hewitt J, Lyon MJ, Marshall JA, McIver CJ, Rawlinson WD, White PA. 2008. Epidemics of gastroenteritis during 2006 were associated with the spread of norovirus GII.4 variants 2006a and 2006b. *Clin Infect Dis* 46: 413-420.

van Asten L, van den Wijngaard C, van Pelt W, van de Kassteele J, Meijer A, van der Hoek W, Kretzschmar M, Koopmans M. 2012. Mortality attributable to 9 common infections: significant effect of influenza A, respiratory syncytial virus, influenza B, norovirus, and parainfluenza in elderly persons. *J Infect Dis* 206: 628-639.

van Beek J, Ambert-Balay K, Botteldoorn N, Eden JS, Fonager J, Hewitt J, Iritani N, Kroneman A, Vennema[H], Vinjé J, White PA, Koopmans M, on behalf of NoroNet. 2013. Indications for worldwide increased norovirus activity associated with emergence of a new variant of genotype II.4, late 2012. Euro Surveill 18: pii=20345

Vega E, Barclay L, Gregoricus N, Williams K, Lee D, Vinjé J. 2011. Novel surveillance network for norovirus gastroenteritis outbreaks, United States. *Emerg Infect Dis*; 17: 1389-1395.

Velasquez LS, Shira S, Berta AN, Kilbourne J, Medi BM, Tizard I, Ni Y, Arntzen CJ, Herbst-Kralovetz MM. 2011. Intranasal delivery of Norwalk virus-like particles formulated in an in situ gelling, dry powder vaccine. *Vaccine* 29: 5221-5231.

Verhoef L, Kouyos RD, Vennema H, Kroneman A, Siebenga J, van Pelt W, Koopmans M; Foodborne Viruses in Europe Network. 2011. An integrated approach to identifying international foodborne norovirus outbreaks. *Emerg Infect Dis* 17: 412-418.

Verhoef L, Hewitt J, Barclay L, Ahmed SM, Lake R, Hall AJ, Lopman B, Kroneman A, Vennema H, Vinjé J, Koopmans M. 2015. *Norovirus genotype profiles associated with foodborne transmission,* 1999-2012. Emerg Infect Dis 21: 592-599.

Verma V, Tan W, Puth S, Cho KO, Lee SE, Rhee JH. 2016. Norovirus (NoV) specific protective immune responses induced by recombinant P dimer vaccine are enhanced by the mucosal adjuvant FlaB. *J Transl Med* 14: 135.

Vinjé J. 2015. Advances in laboratory methods for detection and typing of norovirus. *J Clin Microbiol* 53: 373-381.

Vinjé J, Altena SA, Koopmans MP. 1997. The incidence and genetic variability of small round-structured viruses in outbreaks of gastroenteritis in The Netherlands. *J Infect Dis* 176: 1374-1378.

Vinjé J, Green J, Lewis DC, Gallimore CI, Brown DW, Koopmans MP. 2000. Genetic polymorphism across regions of the three open reading frames of "Norwalk-like viruses". *Arch Virol* 145: 223-241.

Wang J, Jiang X, Madore HP, Gray J, Desselberger U, Ando T, Seto Y, Oishi I, Lew JF, Green KY. 1994. Sequence diversity of small, round-structured viruses in theNorwalk virus group. J Virol 68:5982-5990.

Wang L, Huang P, Fang H, Xia M, Zhong W, McNeal MM, Jiang X, Tan M. 2013. Polyvalent complexes for vaccine development. *Biomaterials* 34: 4480-4492.

Wang L, Cao D, Wei C, Meng XJ, Jiang X, Tan M. 2014. A dual vaccine candidate against norovirus and hepatitis E virus. *Vaccine* 32: 445-452.

Widdowson MA, Cramer EH, Hadley L, Bresee JS, Beard RS, Bulens SN, Charles M, Chege W, Isakbaeva E, Wright JG, Mintz E, Forney D, Massey J, Glass RI, Monroe SS. 2004. Outbreaks of acute gastroenteritis on cruise ships and on land: identification of a predominant circulating strain of norovirus--United States, 2002. *J Infect Dis* 190: 27-36.

Widdowson MA, Rockx B, Schepp R, van der Poel WH, Vinje J, van Duynhoven YT, Koopmans MP. 2005. Detection of serum antibodies to bovine norovirus in veterinarians and the general population in the Netherlands. *J Med Virol* 76: 119-128.

Xia M, Tan M, Wei C, Zhong W, Wang L, McNeal M, Jiang X. 2011. A candidate dual vaccine against influenza and noroviruses. *Vaccine* 29: 7670-7677.

Xia M, Wei C, Wang L, Cao D, Meng XJ, Jiang X, Tan M. 2016. A trivalent vaccine candidate against hepatitis E virus, norovirus, and astrovirus. *Vaccine* 34: 905-913.

Yazawa S, Yokobori T, Ueta G, Ide M, Altan B, Thongprachum A, Nishimura T, Nakajima T, Kominato Y, Asao T, Saniabadi AR, Furukawa K, Kuwano H, Le Pendu J, Ushijima H. 2014. Blood Group Substances as Potential Therapeutic Agents for the Prevention and Treatment of Infection with Noroviruses Proving Novel Binding Patterns in Human Tissues. *PLoS One* 9: e89071.

Yi W, Shao J, Zhu L, Li M, Singh M, Lu Y, Lin S, Li H, Ryu K, Shen J, Guo H, Yao Q, Bush CA, Wang PG. 2005. *Escherichia coli* O86 O-antigen biosynthetic gene cluster and stepwise enzymatic synthesis of human blood group B antigen tetrasaccharide. *J Am Chem Soc* 127: 2040-2041.

Zheng DP, Ando T, Fankhauser RL, Beard RS, Glass RI, Monroe SS. 2006. Norovirus classification and proposed strain nomenclature. *Virol* 346:312-323.

Zheng DP, Widdowson MA, Glass RI, Vinjé J. 2010. Molecular epidemiology of genogroup II-genotype 4 noroviruses in the United States between 1994 and 2006. *J Clin Microbiol* 48: 168-177.

About the Editor

Prof. Jesús L. Romalde, Full Professor of Microbiology since 2010, has been involved in research on enteric viruses for more than 25 years, including different aspects such as bioacumulation, depuration and transmission of norovirus and hepatitis A virus by bivalve shellfish, prevalence in human populations, or viral genotyping, among others. In the last years, he has led investigations on emerging viruses including sapovirus, hepatitis E virus and aichivirus. Since 1987, he has participated in 40 founded R+D projects from public calls, being the principal investigator in 17 of them. In addition, he has been involved in more than 20 industry-funded projects, being the scientist in charge in 10 of them. He has co-authored 211 publications (171 research papers and 40 book chapters and monographies) and has presented 25 invited talks. He has directed 12 Ph.D. thesis and 20 Master research projects. Between 2012 and 2014, he holded the position of Vice-rector for Degrees and Teaching and Research Staff and from January 2017 is the Director of the Department of Microbiology and Parasitology of the University of Santiago de Compostela (Spain).

Index

#

A

B

C

D

E

F

G

I

L

M

N

O

P

Q

R

S

T

V

W

Z